ENGINEERING DESIGN, SECOND EDITION

Contrary to popular mythology, the designs of successful products and systems do not suddenly or magically appear. The authors believe that symbolic representation, and related problem-solving methods, offer significant opportunities to clarify and articulate concepts of design to lay a better framework for design research and design education. Artificial intelligence (AI) provides a substantial body of material concerned with understanding and modeling cognitive processes. This book adopts the vocabulary and paradigms of AI to enhance the presentation and explanation of design. It includes concepts from AI because of their explanatory power and their utility as possible ingredients of practical design activity. This second edition is enriched by the inclusion of recent work on design reasoning, computational design, AI in design, and design cognition, with pointers to a wide cross section of the current literature.

Clive L. Dym is Fletcher Jones Professor of Engineering Design and Director of the Center for Design Education at Harvey Mudd College. His primary interests are engineering design and structural mechanics. Dym is the author of several books, including *Engineering Design: A Synthesis of Views* (1994); *Structural Modeling and Analysis* (1997); *Principles of Mathematical Modeling, Second Edition* (2004); and *Engineering Design: A Project-Based Introduction, Third Edition*, coauthored with Patrick Little (2008). He has published more than 90 refereed journal articles and is the founding editor of *Artificial Intelligence for Engineering Design, Analysis and Manufacturing*. Dym has served on the editorial boards of several other journals, including the American Society of Mechanical Engineers' *Journal of Mechanical Design*. Dym is co-winner of the NAE's 2012 Bernard M. Gordon Prize.

David C. Brown is Professor of Computer Science and has a collaborative appointment as Professor of Mechanical Engineering at Worcester Polytechnic Institute. He is also affiliated with the Manufacturing, Robotics, and Learning Sciences programs. Brown's research interests include computational models of engineering design and the applications of artificial intelligence to engineering and manufacturing. He authored *Design Problem Solving: Knowledge Structures and Control Strategies*, with B. Chandrasekaran (1990), and edited *Intelligent Computer Aided Design*, with M. B. Waldron and H. Yoshikawa (1992). From 2001 to 2011, Brown was Editor in Chief of *Artificial Intelligence in Engineering Design, Analysis and Manufacturing*, and he has served on the editorial boards for *Concurrent Engineering: Research and Application, Research in Engineering Design*, and *The International Journal of Design Computing*.

Engineering Design

REPRESENTATION AND REASONING

Second Edition

Clive L. Dym
Harvey Mudd College

David C. Brown
Worcester Polytechnic Institute

CAMBRIDGE
UNIVERSITY PRESS

CAMBRIDGE
UNIVERSITY PRESS

32 Avenue of the Americas, New York NY 10013-2473, USA

Cambridge University Press is part of the University of Cambridge.

It furthers the University's mission by disseminating knowledge in the pursuit of education, learning and research at the highest international levels of excellence.

www.cambridge.org
Information on this title: www.cambridge.org/9781107697140

First edition published as *Engineering Design: A Synthesis of Views* 1994
Second edition first published 2012
First paperback edition 2014

A catalogue record for this publication is available from the British Library

Library of Congress Cataloguing in Publication data
Dym, Clive L.
Engineering design : representation and reasoning / Clive L. Dym, David C. Brown. – 2nd ed.
   p. cm.
Includes bibliographical references and index.
ISBN 978-0-521-51429-3 (hardback)
1. Engineering design – Data processing.   2. Computer-aided engineering.   I. Brown, David C., 1947–   II. Title.
TA174.D964   2012
620'.004201–dc23   2011041119

ISBN 978-0-521-51429-3 Hardback
ISBN 978-1-107-69714-0 Paperback

We dedicate this book to our brothers,
Harry Dym and Martin Brown

Reflections

Knowledge is an artifact, worthy of design.

– Stefik and Conway (1982)

It has often been said that a person doesn't really understand something until he teaches it to someone else. Actually a person doesn't understand something until he can teach it to a computer, i.e., express it as an algorithm...

The attempt to formalize things as algorithms leads to a much deeper understanding than if we simply try to understand things in the traditional way.

– Knuth (1973)

Contents

Preface

More than 15 years have passed since the publication of *Engineering Design: A Synthesis of Views*. Significant new material has emerged during this time that deserves to be incorporated into the unified view presented in the first edition. In addition, a larger group of people now believe that designing is not just analysis and that research in design offers the possibility of studying a collection of cognitive processes that allows significant insights. Thus, for this new edition, we replace the first edition subtitle with *Representation and Reasoning* to make its focus more explicit.

Our approach was to leave the first edition more or less untouched while adding material that clarifies, updates, and extends the original text. This new material was placed in sidebars that are keyed to appropriate points in the (original) text. Any new references were added to the end of the book to allow students and scholars to explore topics that interest them in more detail. We hope that these additions will extend the life of the book for another 15 years (or even more!).

The first edition leaned heavily on the field of artificial intelligence (AI). AI-based models of design, and the techniques AI brings to bear, have changed since 1994. For example, as the field's understanding of routine and near-routine design strengthened, researchers moved on to consider nonroutine tasks.

Although design education has changed, it has changed slowly. There are still many educators and engineers who would benefit from the lessons of the first edition as well as from the updated material. Engineering design education and design research in general have grown in importance – and it would be nice to think that the first edition has played a constructive role in this growth. There is now a new and wider audience for an expanded and updated "synthesis of views."

We intend our audience to be engineers interested in understanding more about engineering design, designers interested in understanding more about the utility of AI, and AI researchers interested in design. We believe that this book will continue to attract engineers who do not have any prior knowledge or commitment to this updated view of design as an area of research and study.

The authors still believe this statement from the Preface of the first edition:

> ... symbolic representation and related problem-solving methods, offer significant opportunities to clarify and articulate concepts of design so as to lay a better framework for design research and design education. Inasmuch as there is within AI a substantial body of material concerned with understanding and modeling cognitive processes, and because the level of articulation in this work transcends in many ways the common vocabulary of engineering design, we may find it useful to adapt (where possible) and appropriate the vocabulary and paradigms of AI to enhance our understanding of design.

As a consequence, this second edition continues to include concepts from AI not only because of their explanatory power but also because of their utility as possible ingredients of practical design activity. Of course, much has changed in the AI modeling of design and in AI in general, so we have updated the material appropriately. With the growth in cognitively influenced studies of design since the first edition, we have also selectively incorporated material from that field. Our goal is still to provide a focused, concise synthesis, without being so comprehensive that the book becomes cumbersome.

As you can see, we have retained the Preface to the first edition because much of it is still highly relevant, and it still sets the tone for the approach used in this second edition. Some of the updated topics include features, functional representation and reasoning, affordances, design rationale, ontologies, grammars, genetic algorithms, routineness, creativity, assumptions, design decision making, analogy, and collections of agents as representations of teams.

Because we have published widely in this field for many years and share a long history of association with the Cambridge University Press journal *AIEDAM: Artificial Intelligence for Engineering Design, Analysis and Manufacturing*, it should be no surprise that we lean heavily on our papers and this journal in what we cite. We have included a wide variety of the work of others, but we hope you will forgive us for any unintentional bias.

Engineering design education and design research has changed and grown in importance since the first edition. We hope that many more engineers, students, and teachers should now be able to benefit from this updated synthesis of views about representation and reasoning in engineering design.

Clive L. Dym   David C. Brown
Harvey Mudd College   Worcester Polytechnic Institute
Claremont, California   Worcester, Massachusetts

Preface to the First Edition

Design is a central activity in engineering. Indeed, Herbert A. Simon has argued that design is *the* central activity that defines engineering – or, at the very least, distinguishes it from the "pure" sciences – because the role of engineering is the creation of artifacts. And yet, many of us within the engineering community believe that design is a misunderstood activity that is not well represented in engineering education or research. We are very much aware that engineering science dominates the intellectual landscape of engineering today, and it is certainly arguable that analysis dominates both engineering education and research. Indeed, it has long been a concern that design is improperly taught and inadequately represented in engineering curricula and that too often design is seen as legitimate only when it can be explained in terms of analysis (as in the notion that design is "iterative analysis").

One of the problems that design educators and design researchers have faced is a perceived lack of rigor, and this perception has in turn led to calls for a more organized, more "scientific" focus to research about design. One aspect of this perception is that design is viewed as a "soft" subject: design is not a "hard" discipline because it is not sufficiently mathematical. Another aspect is that the vocabulary for analyzing and describing design is not shared, even within the design community, although this situation has begun to improve.

We do not intend to revisit the (familiar) arguments about how this state of affairs has evolved or who, if anyone, is to blame. Instead, we focus on how our understanding of the discipline of design can be broadened and strengthened and on ways in which we can discuss design in a more coherent and precise way. In particular, we will try to demonstrate that recent advances in the field of artificial intelligence (AI), particularly symbolic representation and related problem-solving methods, offer significant opportunities to clarify and articulate concepts of design so as to lay a better framework for design research and design education. Inasmuch as there is within AI a substantial body of material concerned with understanding and modeling cognitive processes, and because the level of articulation in this work transcends in many ways the common vocabulary of engineering design, we may find it useful to adapt (where possible) and appropriate the vocabulary and paradigms of AI to enhance our understanding of design.

We recognize first that the modeling techniques that currently occupy most of the various engineering curricula, rooted in applied mathematics and usually quite adequate for analysis, do not in fact represent a vocabulary complete enough for the synthesis task of generating and choosing among different designs. The situation is perhaps confused because analytical tools are often used to explore particular conceptual designs down through a chain of detailed designs to a final design. What appears to be missing is a language or a means of representing designs more abstractly than is required for detailed design, for example, but with enough structure (e.g., hierarchies and networks) to allow sensible articulation of the issues involved in the validation of a design at their corresponding levels of detail. Furthermore, this vocabulary (or language or, even, set of languages) ought to be recognizable and useful across all engineering disciplines, even if it is not fully applied in every single instance.

Our basic thesis is that recent advances in AI research offer useful prospects for representing the kinds of intelligent, informed design knowledge that is beyond the current scope of mathematical modeling. The single most relevant development of interest here is the idea of *symbolic representation*, which allows in computational terms the processing of lists of words, which in turn facilitates – and even encourages – the representation of objects and their attributes in a fairly general way. Inasmuch as these objects can be conceptual as well as physical, the foundation has been laid for representing qualitative aspects of our thinking about design in ways that we could not achieve heretofore. The vocabulary of this AI-based research offers interesting opportunities for articulating concerns about particular designs and about the field of design itself. That is, recent attempts to use symbolic representation to make design computable have perforce led to an articulation of the design process that should be quite useful to the engineering community in its ongoing examination of design in education, practice, and research.

Thus, we have directed this monograph toward synthesizing an operational definition of engineering design to better articulate what we mean by engineering design, how we can discuss both the design of artifacts and the process of design, and which areas are most amenable to – and perhaps most require – formal research approaches. The central theme of our discussion, that representation is the key element in design, parallels the polymathic vision of Simon, who listed representation as one of the seven subjects in the ideal engineering design curriculum. In addition, akin to the quotation (p. vi) from Knuth, we argue that recent research in AI aimed at rendering design computable has provided new techniques for design representation that enable us to better explicate design concepts and processes. These developments have at the same time given us computer-aided design tools of unparalleled power and flexibility. Another consequence of these recent developments is that we now have available new ways of designing our design knowledge (cf. Stefik and Conway, as quoted on p. vi) and corresponding problem-solving paradigms that should be incorporated into the *weltanschauung* of engineers and designers.

However, although much of this discussion has been motivated by current research aimed at making design (and other cognitive processes) computable in some sense, we most emphatically do not argue that all design should be automated or even made computable. Design activities encompass a spectrum from *routine* design of familiar parts and devices, through *variant* design that requires some modification in form or function, to truly *creative* design of new artifacts. It is difficult to argue that we can at present model truly creative design. However, although we may not be able to model – and thus describe and teach – creativity, we must recognize that the spectrum of design concerns does include many processes that are susceptible to thoughtful analysis – in other words, that are cognitive processes. Traditional designers and teachers of design might well complain about the characterization of design as a cognitive process, but we would respond that this is due to confusion about where creativity and thoughtful process interact and overlap, on the one hand, and where they are distinct, on the other. The boundary between creativity and what we recognize as a cognitive process is a moving one, especially in terms of our understanding, so we must be careful not to develop a new orthodoxy about design that prejudges where that boundary is and where, as a result, we preclude what we can learn and teach about design.

We also do not claim that AI has all the answers. Rather, we believe that AI-based efforts aimed at increasing the options for representing designed artifacts and the design process are helping produce a deeper awareness of what is involved in design as well as a vocabulary for discussing design. It is this awareness and its articulation that is worth exploring and adapting in order to improve the art of engineering. And because design tasks vary from the routine to the creative (or from adaptive through variant to original), we suggest that the focus of research in computer-aided design should not be to simply automate design. Rather, we believe that such research could be viewed as having both a "science" component and an "engineering" component, the former being perhaps more basic than the latter. The goals of the more basic research would include developing better representations, languages, and problem-solving methods as well as a deeper understanding of the kind of knowledge used to solve engineering design (and analysis) problems. The goal of engineering-oriented research could be said to be the creation of *designer's assistants* that support and facilitate the exploration of design alternatives or (perhaps radically) different designs. In this respect, we also point out that such designer's assistants have produced some very tangible gains in product design (cf. Section 1.2).

This book is organized as follows. In Chapter 1, we frame the basic issues in greater detail. In Chapter 2, we review some definitions of design, both in engineering and in other domains, and we present a working definition of engineering design that is sufficiently abstract to acknowledge design both for the production of plans for making artifacts and as a process in itself. We devote the next two chapters to defining and characterizing the design process as a thought process, and we pay special attention to embedding some of the newer ideas into traditional views of design. In Chapter 3, we outline some traditional views of the design process, both descriptive and prescriptive, and then we present some more recent descriptions that reflect some of the research we just mentioned. We then proceed to taxonomies of design (Chapter 4); that is, we try to characterize design tasks and refine them in some detail. Such formalization of the design process has been done only relatively recently, and so we will probably raise more questions about such taxonomies than we will provide answers. An extended discussion of artifact representation is given next (Chapter 5), with particular emphasis on representations used in engineering and buttressed by examples drawn from recent engineering design research results. Chapter 5 also raises questions about communicating design requirements and ideas, particularly in the context of representing design attributes in shared, transmittable ways.

In Chapter 6, we turn to the next logical step, the representation of the design process. We start with a look at some key AI-based problem-solving methods and then review briefly some classical design aids. We do this to introduce some of the vocabulary (and the underlying ideas) used to describe design as a process of articulating and solving engineering problems. Then, having identified a language suitable for discourse about the discipline of design, we provide illustrations of new representations of design knowledge and how these representations help us apply that knowledge in design processes. In Chapter 7, we review some of the current research in engineering design representation with an eye toward identifying future trends. In this final chapter, we also look at the roles that symbolic representation and knowledge-based (expert) systems can play in engineering design, in both practice and education.

In Chapters 5 and 6, especially, we cast many examples in the style of object-oriented programming. In the research and applications literature, these illustrations are presented in pseudocode. Although there are some common approaches (e.g., object names, attributes, values, and procedures are typically written in a sans serif type such as Helvetica), there is also a wide variety of practices that depend on the preferences of individual researchers (or, perhaps more accurately, the capabilities of their software package!). We adopted Helvetica as the preferred font for all such examples, but we follow the preferences of the cited authors when it comes to capitalization, underscores, and so on. Thus, it may appear at first glance that we are being somewhat inconsistent in our presentation. In fact, however, we are striving for consistency with the typeface and the intentions of the works cited.

Finally, a note on referencing the many works whose ideas we build on in this rather personal view of design. In order to not distract you, the reader, with numerous citations in the text, the "bibliographic notes" appear at the end of each chapter. The notes outline particular concepts and ideas, along with their appropriate citations, in what is often called the "social science" style (i.e., author(s) (date)). The notes are organized by chapter sections, and the citations are keyed to the reference list found at the end of the book. We worked very hard to be both complete and fair as we compiled these notes and citations. However, perhaps flaunting our humanity, we apologize in advance for any errors in this regard and ask for a divine response from those who have been inadvertently forgotten or improperly cited!

Acknowledgments

From the First Edition

Steven J. Fenves (Carnegie Mellon University) provided an encouraging ear as I talked my way through a very early vision of this book, and he has maintained a friendly interest ever since.

Daniel R. Rehak (Carnegie Mellon University) exercised his unique critical skills intelligently and perceptively over much of the life of the project. He did his best to keep me intellectually honest. What more could one ask of a critic?

Dr. Conrad Guettler and Ms. Florence Padgett of Cambridge University Press have also earned large measures of gratitude. For some years, Conrad gently but persistently encouraged me to write this monograph, and Florence provided cheerful and unflagging editorial support during my life as a Press author.

Erik K. Antonsson (California Institute of Technology) and David M. Steier (Price Waterhouse) provided helpful comments on some of the early chapters of the book. So did my biking buddy, Alicia Ross of Claremont, who listened to me ramble as we cycled around the Inland Empire.

Several publishers, corporations, and individuals have kindly granted me permission to reproduce or replicate figures that have originally appeared elsewhere. The citations for the particular works from which figures have been taken or adpated are given in the respective figure captions. Thus, my sincere thanks to Addison-Wesley (5.11–5.14); American Society of Civil Engineers (5.18–5.21); American Society of Mechanical Engineers (4.1, 4.2, 5.4, 6.12); Butterworth-Heinemann (5.9, 5.15–5.17); John R. Dixon (5.5, 5.6); Steven J. Fenves (1.1); James H. Garrett, Jr. (5.22); R. Gronewold, C. White, C. Nichols, M. Shane, and K. Wong (6.1); B. Hartmann, B. Hulse, S. Jayaweera, A. Lamb, B. Massey, and R. Minneman (6.2, 6.3, 6.6); Institute of Electrical and Electronics Engineers (6.11, 6.15, 6.18, 6.19); International Federation for Information Processing (5.3, 6.13, 6.14); Macmillan (3.1); McGraw-Hill (6.7–6.10); Pergamon Press (3.4, 5.2); Southern California Edison (5.1, 5.10); Springer-Verlag (3.2, 4.3–4.8); John Wiley (3.3, 6.4, 6.5); and Xerox Corporation (5.7, 5.8, 6.16, 6.20).

For the Second Edition

Clive wants to thank Joan Dym for her patience, understanding, and support as he worked to complete yet another book, and Dave wants to thank his family, friends, and coworkers. We both would like to thank our editor, Peter Gordon, for his encouragement and optimism.

1 Framing the Issues

The principal thesis of this book is that the key element of design is *representation*. If we were to consult a standard dictionary, we would find *representation* defined as "the likeness, or image, or account of, or performance of, or production of an artifact." Note, however, that whereas our dictionary defines representation as a *noun* in which terms such as *image* and *likeness* refer to the artifact being designed, it also suggests aspects of a *verb* when it defines the design process in terms of a performance or a production. This suggests that representation in design incorporates both representation of the *artifact* being designed as well as representation of the *process* by which the design is completed. We now examine briefly both types of representation.

1.1 Representation of Artifacts for Design

Suppose we are charged with the design of a safe ladder. What does it mean, first of all, for a ladder to be "safe"? That it should not tip on level ground? That it should not tip on a mild slope? What is a mild slope? How much weight should a safe ladder support? Of what material should it be made? How should the steps be attached to the frame? Should the ladder be portable? What color should it be? How much should it cost? Is there a market for this ladder?

We have quickly identified several – but by no means all – of the questions in this very simple design problem, and we would not be able to answer most of them by just applying the mathematical models that originate in the engineering sciences. For example, we could use Newton's equilibrium law and elementary statics to analyze the stability of the ladder under given loads on a specified surface, and we could write beam equations to calculate the bending deflections and stresses of the steps under the given loads. But which equations do we use to define the meaning of "safe" in this context or to define the color or the marketability? In fact, which equations do we use to describe the basic function of the ladder? We know that the function the ladder serves is to allow someone to climb up some vertical distance, perhaps to paint a wall, perhaps to rescue a cat from a tree limb, but it is for the designer to translate these verbal statements of function into some appropriate mathematical models at some appropriate time – often early on – in the design process. That is, we have no mathematical models that describe function directly; we infer functional behavior by reasoning about results obtained from manipulating mathematical models.

Thus, we recognize already that a *multiplicity or diversity of representations* is needed for design, a collection of representation schemes that would enable description of those issues for which analytical physics-based models are appropriate; those that require geometric or visual analysis to reason about shape and fit; those that require economic or other quantitative analysis; and those requiring verbal statements not easily expressed in formulas or algorithms. Some of the verbal requirements could be statements about *function*; about *form* (e.g., a stepladder or an extension ladder); about *intent* (e.g., to be used at home or to be used in an industrial environment); or about legal requirements (e.g., to satisfy government regulations).◆

Verbal statements are also made to describe or reference *behavior* (e.g., steps on the ladder should not have too much give), *heuristics* (e.g., my experience suggests that fiberglass ladders feel stiffer than aluminum ladders), *design decision alternatives* (e.g., Can I choose between a stepladder and an extension ladder?), *preferences* (e.g., I like ladders that are bright blue), *affordances* (e.g., steps spaced in small, uniform increments enable climbing), *constraints* (e.g., the ladder can cost no more than $100), *assumptions* (e.g., I thought it would fit in my trunk), and *intent* or *rationale* (e.g., the ladder is for home use).

In essence, representation is modeling. However, representation in design is much broader than modeling in engineering science, wherein mathematical modeling is the key idea. In fact, a more apt analogy may be found in the linguistic notion of abstraction ladders or in Korzybski's aphorism, "The map is not the territory." The real point, however, is that we must represent meaningfully much more knowledge than can be set into mathematical formulas or numerical realizations of those formulas, and this is now possible. Advances in computing generated by AI research allow – and even encourage – the representation of symbols and thus objects, attributes, relationships, concepts, and so on. New programming styles have emerged in which we can capture more abstract conceptual and reasoned design knowledge that cannot be reduced to conventional algorithms.

Although we have discussed the role of representation in design before, we are not the first nor will we be the last to stress the importance of representation. As we noted previously, representation is one of the seven subjects in Simon's ideal design curriculum. (The other six subjects in the curriculum, in which Simon describes design as the science of the artificial, are evaluation theory, algorithms and heuristics for identifying optimum and satisfactory designs, formal logic for design, heuristic search, resource allocation for search, and the theory of structure and design organization.) A brief sampling of recent design research in which representation figures prominently includes work on features in mechanical design; shape grammars; object-oriented data structures; interaction of form, function, and fabrication; formal theories of design; shape emergence; and so on. We see from this list that the line between representing artifacts and representing the design process is not a sharp one. We will have a chance to explore some of this and related work in Chapters 5 and 6.

It is also important that we recognize that representation is not an end in itself but rather a means to an end; it is a way of setting forth a situation or formulating a problem so that we can find efficiently an acceptable resolution to a design problem. This also implies that representation is strongly coupled to whatever strategy we have chosen for solving a problem (whether in design or in some other domain). Because it is pointless to invoke alternate representations unless we gain some leverage thereby, the notion of changing the representation of a problem is inexorably linked to the idea that there is a problem-solving strategy available to us that uses this representation in a beneficial way. This is not to say that the research objective of developing new representations should be limited to those for which a problem-solving strategy is available. Research on both artifact representation and problem solving should proceed independently, although perhaps in parallel. But the development of new representations does suggest that broader paradigms of problem solving should be integrated into the outlook of engineers and designers, the idea being that approaches such as AI-based paradigms and tools will become part of the arsenal of weapons available for better engineering.

1.2 Representation of the Design Process

Let us return to the ladder-design problem, now with a view toward examining the process by which the design will be done. First, we recognize that the initial statement of the client's wishes is rather vague, in large part because it is simply a brief verbal description. In fact, design projects often originate with a brief verbal statement, such as President John F. Kennedy's lunar challenge. To proceed with a design, we have to flesh out these skimpy skeletons by *clarifying* and *translating* the client's wishes into more concrete objectives toward which we can work. In the clarification step, we ask the client to be more precise about what is really wanted by asking her or him questions: For what purposes is the ladder to be used? Where? How much can the ladder itself weigh? What level of quality do you want in this ladder? How much are you willing to spend? However, the degree of precision that we might demand from the client could well depend on where we are in the design process.

Some of the questions we asked in the hope of clarifying the client's wishes obviously connect with our previous discussion (cf. Section 1.1) on artifact representation, but some lead us into a process in which we begin to make choices, analyze the dependencies and interrelationships between possibly competing choices, assess the trade-offs in these choices, and evaluate the effect of these choices on our overall goal of designing a safe ladder. (There are formal methodologies for identifying trade-off strategies.) For example, the form or configuration of the ladder is strongly related to its function: We are more likely to use an extension ladder to rescue a cat from a tree and a stepladder to paint the walls of a room. Similarly, the weight of the ladder will certainly have an impact on the efficiency with which it can be used to achieve its various purposes: aluminum extension ladders have replaced wooden ones largely because of the difference in weight. The material of which the ladder is made not only influences its weight; it also is very influential in determining its cost and even its feel. Wooden extension ladders are considerably stiffer than their aluminum counterparts, so users of the aluminum versions have to get used to feeling a certain amount of "give" and flex in the ladder, especially when it is extended significantly. Thus, a possible design goal that was not even mentioned has suddenly emerged: design a safe, *stiff* ladder.

In the translation step, we convert the client's wishes into a set of *design specifications* that serve as benchmarks against which to measure the performance of the artifact being designed. The translation process is where the "rubber begins to meet the road," for it is here that the verbal statement is recast in terms of more specific design objectives. These specific objectives can be stated in a number of ways, reflecting variously the desire to articulate specific dimensions or other attributes of the designed object, which are usually called *prescriptive specifications*; specific procedures for calculating attributes or behavior, which are embedded in *procedural specifications*; or the desired behavior of the device, which is encoded in *performance specifications*. A successful design is one in which performance meets (or exceeds) the given specifications and satisfies (or exceeds) the client's expectations. We reiterate that the specifications may evolve or be further detailed and refined as the design unfolds. The role of specifications in design has been the subject of much thought and discussion, and it is also clear that the techniques for stating design specifications (and, later, fabrication specifications) are intimately related to design-representation issues.

The design process is evolutionary in nature, and we will come across choices to make and different paths to follow as a design unfolds. In fact, the particular choices that present themselves often become evident only after we have refined the original design objective – the client's statement – to some extent. For example, at some point in our ladder design, we have to confront the issue of fastening the steps to the ladder frame. The choices will be influenced by the desired behavior (e.g., although the ladder itself may flex somewhat, it would be highly undesirable for the individual steps to have much give with respect to the ladder frame) as well as by manufacturing or assembly considerations (e.g., would it be better to nail in the steps of a wooden ladder, or use dowels and glue, or perhaps nuts and bolts?).♦

> Sometimes (e.g., in Gero's (1990) simple and often-cited model of design), the design process is driven by comparison between the design's expected behavior, derived from the desired function, and the predicted or actual behavior, resulting from the design's structure.

Thus, the choices themselves need to be articulated in some language naturally conducive to making them; that is, choosing a particular bolt and nut pair to achieve a certain fastening strength requires access to a manufacturer's catalog as well as to the results of calculations about bearing and shear stresses. The particular process involved here could be called *component selection*, and it is invoked after we have *decomposed* the form of the ladder into its components or pieces, and after we have selected a particular type of component.

We are using this very simple example to illustrate the *formalization* of the design process through which we make explicit the ways we are doing some elements of our design. We could say we are *externalizing* aspects of the process so that we can move them from our heads into some recognizable language(s) for further analysis. There is no shortage of attempts to externalize design engineering processes, and we review many of these process models in Chapter 3. These descriptions and prescriptions are externalized to the extent that we can draw flow charts to describe the major steps of a design process, and the descriptions do point to analyses that need to be done and choices that must be evaluated, some of which can be done with conventional algorithms. However, these descriptions cannot be made computable because they are all relatively abstract; that is, they are not refined enough or rendered in sufficient detail that we can identify the underlying thought processes. Again, the objective in refining these processes is not just to be able to render them computable; it is to be able to *analyze* them in sufficient detail that we can *synthesize* design processes out of their fundamental constituent processes. When we do so in earnest in Chapter 6, we will see that we are taking advantage of research in AI (and related fields such as cognitive science) to examine and describe the activity that is called *design*. We view this as the *representation of the design process* as opposed to the representation of the artifacts that are being designed.

A recent knowledge-based system that illustrates the capture of a design process is called PRIDE; it serves as a designer's assistant for the mechanical design of paper-handling subsystems in copiers (see Chapter 6). Designing paper-transport systems for copiers is difficult because of the number and kind of design variables and their complex interactions. Nonetheless, by identifying the way designers actually do this task, the designers of the PRIDE system built a knowledge-based system that does much of the same design task as human designers. That is, PRIDE uses a variety of representation formalisms to incorporate both algorithmic and heuristic aspects of the design problem. It also uses a variety of *inference schemes* (i.e., reasoning patterns) and a powerful graphics interface to achieve a relatively complete simulation of the way human designers actually design paper-handling subsystems for copiers. The PRIDE environment allows the designer to experiment with different designs, both graphically and procedurally, and it facilitates the tracking of dependencies between design decisions and the maintenance of multiple design paths. The PRIDE system replicates a designer's approach to a complex problem in a way that simply cannot be done in a conventional, numerically based algorithm.

Furthermore, the PRIDE system works so well that it allows experienced designers to do feasibility studies in just a few hours, whereas it used to take four weeks to develop similar designs. In addition, the PRIDE system is viewed as useful because it also has led to paper-copier designs that are both more consistent and of higher quality.

We note two related points. First, the kind of replication or modeling of a design process that is found in the PRIDE system cannot be achieved by simply extending the traditional engineering science approaches to incorporate the thinking and logic characteristic of operations research (OR), as has been implied by some. The reason for this is that the representations inherent in OR approaches, although they permit the inclusion of economic or similar performance metrics, do not admit those qualitative or strategic choices that cannot be reduced to numbers.◆

A similar critique can be made about the use of *decision theory* (referred to as *decision-based design* (DBD)) because it requires knowledge of probabilities and utilities (see the Decision-Based Design Open Workshop Web pages (DBD 2004)).

The second point is that researchers in other engineering domains (recall that PRIDE's domain is mechanical engineering) have also clearly recognized the utility that knowledge-based (expert) systems have for modeling many phases of the design process – for example, in chemical engineering.

The second line of argument supportive of what has been outlined is that whereas much of the work in design is empirical in nature, both in design practice and in design research, there is apparently no objective basis for describing and evaluating experiments in design. Much of what is known and transmitted about *how* to design artifacts is – or is perceived to be – anecdotal in nature. To the extent that design knowledge is viewed as design *lore*, both the development of the discipline of design and its acceptance by the engineering community as a serious discipline with a rigor and logic of its own are inhibited. Thus, in this context as well, it could prove useful to adopt the relevant terminology and paradigms from AI and related cognitive fields, subjects that are themselves highlighted by experimentation and empirical development. One example is the technique of protocol analysis, which may be described as the process of organizing, understanding, and modeling verbal reports and analyses. This technique has been applied formally and informally to elicit and organize the knowledge that designers use in their own domains. The use of a formal structure and methodology in this particular context is bound to be beneficial in developing a communicable understanding of the process of design.

1.3 An Illustration from Structural Engineering

To illustrate the importance of representation in design and the diversity of representations that we actually use for artifact and process representation, we present now a brief discussion of the structural engineering problem. In essence, the problem is as follows (Figure 1.1). A structural need is identified, whether it is for a mill building or a concert hall. Then we choose a structural concept, perhaps a simple steel frame and steel roof truss for the mill building, something considerably more complex for the concert hall, and we move to preliminary design. In this stage, we usually restrict our efforts to rough sizing of the principal structural members, the object being to see whether the type of structural system that we have chosen is practically feasible. We then move on to flesh out the structure by estimating the types and sizes of the remaining members (e.g., in the mill building, purlins for the roof truss and floor joists as needed). Then we home in on the final, detailed design in which we calculate actual dimensions and placements for all members and their connections. In the final step, we check to ensure that our design meets all statutory requirements, including both applicable building codes and design codes

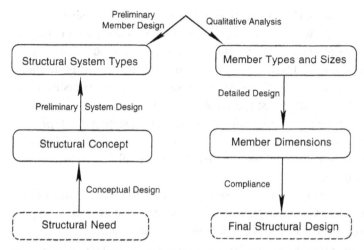

Figure 1.1. A pictorial view of the structural engineering problem (Fenves, 1993).

such as that of the American Institute of Steel Construction (AISC), which lays out performance specifications for steel members and connections.

Let us now examine the kinds of design knowledge deployed in completing such a structural design. Among the kinds of knowledge we apply are classical mechanics (e.g., Newton's laws); structural mechanics (e.g., models of columns and beams); geometry of structures (e.g., relating the geometry of members and assemblages of members to the orientation of the loads they are expected to carry); structural-analysis techniques (e.g., moment distribution for frames and the method of sections for truss analysis); behavioral models (e.g., modeling the stiffness of a complete frame); algorithmic models of structures (e.g., finite element method (FEM) computer codes); structural design codes (e.g., the AISC code); heuristic and experiential knowledge, both derived from practice and encoded in specifications; and *meta-knowledge* about how and where to invoke the other kinds of knowledge. Much of this knowledge is multilayered. For example, our understanding of the behavior of structural systems is realized at three distinct levels: spatial layout (e.g., where to place columns to achieve clear floor spans), functional (e.g., how to support different kinds of loads), and behavioral (e.g., estimating the lateral stiffness of frames).

How do we represent these different kinds of structural design knowledge? In fact, we use several different kinds of representations of the knowledge itself, including mathematical models for classical and structural mechanics (e.g., partial differential equations and variational principles); case-specific analyses (e.g., buckling of slender columns); phenomenological, "back-of-the-envelope" formulas (e.g., the beam-like response of tall buildings); numerical programs (e.g., FEM codes); graphics and computer-aided design and drafting (CADD) packages; rules in design codes (e.g., the AISC code); and heuristic knowledge about structural behavior, analysis techniques, and so forth. Such qualitative knowledge is often subjective and frequently expressed in rules. Thus, we already employ several different representations or "languages" of knowledge, including verbal statements, sketches and pictures, mathematical models, numerically based algorithms, and the heuristics and rules of design codes. When we use these different languages now, we manage to choose (in our head) the right one at the right time; however, in computational terms, we should recognize that it would be desirable to link these different representations or "languages" so that we could model our design process in a seamless fashion. We should also recognize that we often cast the same knowledge in different languages, depending on the immediate problem at hand. For example, a statement (typical of that found in building codes) that the deflection of a floor in a residential building should not exceed its length (in feet) divided by 360 is actually a restatement of equilibrium for a bent beam.

The point we want to make with this example is that for "real" engineering design problems – although it is equally true of our "toy" problem of ladder design – we are already accustomed to handling very complex representation issues. What is beginning to be true now is that we want to formally recognize this in the increasingly elaborate computer-based design tools we are developing. And, even more important for our present purpose, as we try to externalize our design knowledge, we are increasingly conscious of how we think about design. It is this raised consciousness we seek to expose here.

1.4 On the Role of Computation

The final argument we make in this book is, comparatively speaking, relatively straightforward. The rapid advances in the field of computer science, in both software and hardware, have brought increasing opportunities – and pressures – to "computerize" and automate engineering practice as much as possible and, at the very least, to automate the tedious and repetitive parts of engineering.♦

> Current computational resources provide design support by also allowing complex simulations of forces and flows, visualizations of those data, animations of structural models, virtual manufacturing, and rapid prototyping.

We take it as obvious that there are different opportunities for automation in different domains and for different tasks and task types within domains. For example, it has been easier to develop knowledge-based (expert) systems to perform *derivation* tasks, in which assessments are derived from data, than *formation* tasks, in which we attempt to form results to meet specified goals. Similarly, in exploring applications of AI techniques to design, there are going to be differences that ought to be acknowledged from the outset. For example, truly routine design (which is essentially a repetitive process) is much more readily automated than nonroutine design, in which the form and function (or their attributes) of a successful design may not be easily described, if at all. Thus, replication of routine design will offer different opportunities for automating with AI techniques than will the modeling of creative or original design. That is, it is likely that over time, a hierarchy of design tools will be developed to reflect these differing design tasks.

However, the perception of what may be automated – as opposed to what may be encapsulated in a designer's assistant – should not be perceived in static terms. As we articulate and acquire design knowledge, which we must do before we can represent it, we also acquire a keener understanding of that knowledge.♦ This results in a new consciousness of that knowledge, which in turn lays the foundation for discovering new algorithms, new procedures and strategies, or even new representations that may allow more of the process to be automated. Furthermore, as we noted earlier, the boundary between what we can understand and model as a cognitive process and true creativity is a shifting one, and we should not at this point preclude any endeavor that might prevent us from moving that boundary closer to the edge of complete understanding.♦ Still, the goal is not automation of the entire design process; it is the automation of the routine and the boring, and the creation of computer-based tools that facilitate design exploration.

The R1/XCON (McDermott 1982) and PRIDE (Mittal and Dym 1985; Mittal et al. 1986) projects were among the first to note that the knowledge-acquisition process caused knowledge to be articulated that had not been previously recorded (i.e., the acquisition process was worthwhile even without the resulting configuration system). In addition, it is worth noting that an attempt to program the R1/XCON configuration design process as an algorithm failed, whereas using an AI-based technique (rules) for representing knowledge was successful.

Recent research has focused on computational design creativity, arguing that creativity is not a mystery and that it can be studied scientifically and investigated computationally (Boden 1994; Brown 2008).

A final note on computation. We have argued that the mathematics that we use to describe and analyze many engineering problems is inadequate for describing and analyzing many attributes of designed artifacts and design processes. Thus, we need to augment our mathematical modeling tools with others, such as graphics, logic, grammars, word and document processors, and – most relevant to this discussion – those tools based on symbolic representation. We must caution, however, that we are *not* saying that there is no mathematical foundation underlying the symbolic-representation techniques the use of which we advocate. Indeed, there are very complex mathematical problems involved in computation in general and in developing the underlying structure of the kinds of AI programs that are used to develop the kinds of results that we will see later in this book. However, the mathematics involved there is concerned with representing the symbols and the processes used to compute with these symbols so that, ultimately, the computer can do as it is told. Perhaps a very loose analogy is that this particular kind of mathematics is to the symbolic representation that we espouse as set theory and functional analysis are to the continuous mathematical models we routinely employ (e.g., the partial differential equation governing the bending deflection of a plate). Thus, we view as parallel the descriptive representations offered by continuous mathematical models and by symbolic representation of physical and conceptual objects and their attributes and dependencies.

1.5 Bibliographic Notes

The dictionary definition of *representation* is from Woolf (1977).

Section 1.1: Mathematical modeling is discussed in Dym (1983) and Dym and Ivey (1980).♦ Korzybski is quoted in Hayakawa (1978). The new, AI-based programming styles are thoroughly described in Bobrow and Stefik (1986) and Stefik and Bobrow (1985). We have discussed representation before in Dym (1984) and Dym (1992b).♦ Simon's ideal design curriculum is outlined in the classic Simon (1981).♦ Recent design research topics include features in mechanical design (Dixon 1988; Dixon, Cunningham, and Simmons 1989); shape grammars (Stiny 1980); object-oriented data structures (Agogino 1988a); interaction of form, function, and fabrication (Rinderle 1985; Rinderle et al. 1988); formal theories of design (Fitzhorn 1988; Stiny 1988a); and shape emergence (Gero and Yan 1993). A good snapshot of modern AI-based research is Tong and Sriram (1992a, 1992b).

> Mathematical modeling also is discussed by Dym (2004).

> For a good text about knowledge representation and reasoning, see Brachman and Levesque (2004).

> Recent textbooks that suggest material for inclusion in a design curriculum include Dym et al. (2009), Ullman (2009), and Ulrich and Eppinger (2007). Design education continues to be a subject of discussion and research: for example, see Dym et al. (2005), the Web pages of the biennial series of Mudd Design Workshops (MDWs), as well as special issues of the *Journal of Mechanical Design*, "Design Engineering Education" (Doepker and Dym 2007), and *AIEDAM: Artificial Intelligence for Engineering Design, Analysis and Manufacturing*, "Design Pedagogy: Representation and Processes" (Frey et al. 2010).

Section 1.2: Wood and Antonsson (1989, 1990) discuss the role of precision in design. Otto and Antonsson (1991) present a formal methodology for identifying trade-off strategies. The role of specifications in design is discussed in Fenves (1979); Stahl et al. (1983); and Wright, Fenves, and Harris (1980). Techniques for stating design specifications and their relationship to design-representation issues are described in Dym et al. (1988); Garrett and Fenves (1987, 1989); and Garrett and Hakim (1992). "Externalized" models of the design process are given, for example, in Dixon (1966), Woodson (1966), Jones (1970), Pahl and Bietz (1984), Cross (1989), French (1992), and Ullman (1992a, 1992b). Discussions of the PRIDE system include the original paper (Mittal, Dym, and Morjaria 1986) and a retrospective view (Morjaria 1989); see also Section 6.3 and its citations. An interesting view of the role of operations research in design is that of Wilde (1988). Applications of knowledge-based systems to chemical engineering are offered in Lien, Suzuki, and Westerberg (1987). Protocol analysis is defined in Ericsson and Simon (1984); its application to design has been explored formally in Stauffer, Ullman, and Dietterich (1987) and Ullman, Dietterich, and Stauffer (1988) and informally in Mittal and Dym (1985).

Section 1.3: A steel design code can be found in AISC (1986). The types of knowledge deployed in structural engineering are discussed in Dym and Levitt (1991b).

Section 1.4: The ease of building knowledge-based systems is connected to the task they model by Bobrow, Mittal, and Stefik (1986); Dym (1987); and Fenves (1982). Knowledge acquisition for expert systems is discussed by Mittal and Dym (1985). Touretzky (1986) derives the basis for the inheritance structure that underlies the object-oriented descriptions described in Chapter 5.

2 Engineering Design

Design is a ubiquitous word: We see it often and in many different contexts. For example, just in perusing our daily newspapers, we read about people who are automobile designers, dress designers, architectural designers, sound-system designers, aircraft designers, organization designers, highway designers, system designers, and so on and so forth. In fact, design has been a characteristic of human endeavor for as long as we can "remember" or, archaeologically speaking, uncover. Design was done in very primitive societies for purposes as diverse as making basic implements (e.g., flint knives) to making their shelters more habitable (e.g., the wall paintings found in primitive caves). However, because people have been designing artifacts for so long and in so many different circumstances, is it fair to assume that we know what design is, and what designers do and how they do it?

2.1 From Design to Engineering Design

Well, we do know some of the story, but we do not yet know it all. And, of course, one of the themes of this book is that we are still struggling to find ways of externalizing and articulating even that which we do know about design. For example, with regard to the design of elementary artifacts, it is almost certainly true that the "designing" was inextricably linked with the "making" of these primitive implements – that there was no separate, discernible modeling process. However, we can never know for sure, because who is to say that small flint knives, for example, were not consciously used as models for larger, more elaborate cutting instruments? Certainly people must have *thought* about what they were making because they recognized shortcomings or failures of devices already in use and evolved more sophisticated versions of particular artifacts. Even the simple enlargement of a small flint knife to a larger version could have been driven by the inadequacy of the smaller knife for cutting into the hides and innards of larger animals.♦ But we really have no idea of how these early designers thought about their work, what kinds of

A fairly common idea is that reasoning, and also learning, can be failure driven; that is, design reasoning "fails" because a design concept or decision is inadequate in some way. Kant (1985) writes

(continued)

12

languages or images they used to process their thoughts about design, or what mental models they may have used to assess function or judge form. If we can be sure of anything, it would be that much of what they did was done by trial and error (a process that has a modern reincarnation in the method called *generate and test* in which trial solutions are generated by some unspecified means and then tested against given evaluation criteria).

What we do know about design problems in general – and engineering design problems in particular – is that they are typically *open-ended* and *ill-structured*, by which we mean the following:

> (*continued*)
>
> about portions of a designer's refinement process being "difficulty driven," where such difficulties include missing things and inconsistencies, leading to the use of different strategies. In the Soar cognitive architecture, a new goal is created when an "impasse" is reached (i.e., when problem solving cannot continue) (Laird et al. 1993).
>
> Petroski (1992) makes a similar point at a higher level of abstraction, arguing that failure is a major driver for design and that the failure of various devices or systems has spurred efforts to discover more robust engineering solutions.

1. Design problems are said to be *open-ended* because they usually have many acceptable solutions. The quality of uniqueness, so important in many mathematics and analysis problems, simply does not apply.
2. Design problems are said to be *ill-structured* because their solutions cannot normally be found by routinely applying a mathematical formula in a structured way.

It is these two characteristics in particular that make design such a tantalizing and interesting subject. Even the simple ladder design that we broached in Chapter 1 became a complex study because there were choices to make about the form and structure of the solution and because there was no single language for or ordering of the many different design issues that had to be addressed. How much more complicated and interesting are projects to design a new automobile, a skyscraper, or a way to land someone on the moon.

We can find discussions of design well back in recorded history. One of the most famous is the collection of works by the Venetian architect, Andrea Palladio (1508–1580), whose works were apparently first translated into English in the eighteenth century, in which language they are still available today. Discussions of design have been prompted by the concerns of domains as diverse as architecture, decision making in organizations, and styles of professional consultation, including the practice of engineering.

Thus, it should not surprise us that there are many, many definitions of design or that these definitions, at a sufficiently high level of abstraction, seem very much alike. For example, one might say that design is a goal-directed activity, performed by humans, and subject to constraints. The product of this design activity is a plan to realize the goals. Simon offers a definition that seems more closely related to our engineering concerns: the objective of the design activity is the production of a "description of an artifice in terms of its organization and functioning – its interface between inner and outer environments."

At a level intermediate to the definitions just given, Winograd and Flores place

> design in relation to systematic domains of human activity, where the objects of concern are formal structures and the rules for manipulating them. The challenge posed here for design is not simply to create tools that accurately reflect existing domains, but to provide for the creation of new domains. Design serves simultaneously to bring forth and to transform the objects, relations and regularities of the world of our concerns.

This concept of design seems to combine the idea of design as an activity with the explicit articulation of the fact that some objects and their contexts are being represented – and perhaps changed (or created) by manipulating these representations – in order to produce a design. This definition thus seems to parallel the view of representation outlined in Chapter 1.

One aspect of design that we will not explore here, as we move toward a definition of engineering design, is the role of aesthetics in design.[♦] This is largely because the description, representation, and evaluation of aesthetics comprise an extraordinarily difficult task, although suggestions have been made for an algorithmic structure for aesthetics in criticism and design. By excluding aesthetics from further consideration, we do not intend to suggest that aesthetics is unimportant in engineering design; however, we do wish to keep our discussion bounded and manageable.

> It is worth noting that there is now a series of computational aesthetics conferences (with a strong art/graphics flavor) as well as an international society (with a strong mathematics flavor). A brief history of the computational study of aesthetics is provided by Greenfield (2005).

We noted earlier that primitive designers almost certainly designed artifacts through the very process of making them. It has been said that this approach to design is a distinguishing feature of a *craft*. With this in mind, it is also useful for us to distinguish engineering design from those domains – or crafts – wherein the designer actually produces the artifact directly, such as graphics or type design. An engineering designer typically does not produce an artifact; rather, he or she produces a set of *fabrication specifications* for that artifact. That is, the designer in an engineering context produces a detailed description of the designed device so that it can be assembled or manufactured. With the "designing" thus separated from the "making" of the artifact, a host of issues presents itself, although all of the issues can be subsumed in a single statement: the fabrication specification produced by the designer must be such that the fabricator can make the artifact in question without reference to the designer herself or himself. Thus, that specification must be both complete and quite specific; there should be no ambiguity and nothing can be left out.

Traditionally, fabrication specifications have been presented through some combination of drawings (e.g., blueprints, circuit diagrams, and flow charts) and text (e.g., parts lists, materials specifications, and assembly instructions). Although completeness and specificity can be had with these traditional means, we cannot capture a designer's intent in them, which can lead to catastrophe.[♦] The 1981 collapse of Kansas City's Hyatt Regency Hotel occurred because a contractor, unable to procure threaded rods sufficiently long to suspend a second-floor walkway from a roof truss, hung it instead from a fourth-floor walkway using shorter rods. The supports of the fourth-floor walkway were not designed

> Research in design rationale (DR) capture, representation, and use is concerned with recording the designer's intent – that is, the reasons behind the decisions that led to the final design (Burge and Brown 1998). DR has been studied for engineering (Burge and Bracewell 2008) as well as for software engineering (Burge et al. 2008).

to carry the second-floor walkway in addition to its own dead and live loads, so a major disaster ensued. Had the designer been able to signal to the contractor his intentions of having the second-floor walkway suspended directly from the roof truss, this accident might never have happened.

This story also can be used to learn another lesson that is a consequence of the separation of the "making" from the "designing." Had the designer been able to talk with a fabricator or a supplier of threaded rods while the design was still in process, he would have learned that no one manufactured that rod in the lengths needed to hang the second-floor walkway directly from the roof truss. Thus, while still in an early design stage, the designer would have had to seek another solution. It has been true generally in the manufacturing sector that there was a "brick wall" between the design engineers and the manufacturing engineers. Only recently has this wall been penetrated – even broken – and manufacturing and assembly considerations are increasingly addressed while a product is still being designed, rather than afterward. This new practice, called *concurrent engineering*, is becoming very important in design engineering, and it is a field that relies heavily on recent developments in computation, including the integration of multiple representations of artifacts.

The Hyatt Regency tale and the implications we have drawn suggest that a major issue in engineering design – as distinguished from craftlike design – is the clear need for formalisms that can be used to represent fabrication specifications. In designing a formalism for fabrication specifications, we can learn from this disaster that there are requirements of completeness, specificity (or absence of ambiguity), and functional intent that must be addressed. Furthermore, as long as we are developing a new specification formalism, we should require that it also enable us to evaluate a design in terms of how well it meets the original design goals. We thus have another indication of the importance of representation as an issue in engineering design, although our emphasis would have to focus on translating the original design objectives (and constraints) into some version of the specification formalism and on recognizing that these specifications provide the starting point for some manufacturing process. Mostow's definition of design reflects some of these concerns:

> The purpose of design is to derive from a set of specifications a description of an artifact sufficient for its realization. Feasible designs not only satisfy the specifications but take into account other constraints in the design problem arising from the medium in which the design is to be executed (e.g., the strength and properties of materials), the physical

environment in which the design is to be operated (e.g., kinematic and static laws of equilibrium) and from such factors as the cost and the capabilities of the manufacturing technology available.

As we move closer to a definition of engineering design, we should also account for the notion of design as a human activity or process, with all that is thus entailed about context and language. At the same time, as noted before, we must focus on the idea that plans are going to be produced from which an artifact can be realized. Thus, a definition of engineering design must be broad enough to encompass a variety of concerns but not so abstract as to have no obvious practical implementation.

2.2 A Definition of Engineering Design

We adopt as our definition of engineering design the one articulated by Dym and Levitt because it seems to properly situate our exposition:[♦]

> Engineering design is the systematic, intelligent generation and evaluation of specifications for artifacts whose form and function achieve stated objectives and satisfy specified constraints.

A somewhat expanded version of this definition of *design* was proposed by Dym et al. (2005): "*Engineering design* is a systematic, intelligent process in which designers generate, evaluate, and specify concepts for devices, systems, or processes whose form and function achieve clients' objectives or users' needs while satisfying a specified set of constraints" (p. 104).

This definition incorporates many implicit assumptions, some of which we have already anticipated. We now explore the underlying assumptions in some detail, with the express intent of exposing the role of representation in design.

1. As a process, *design is thoughtful and susceptible to understanding*, even if complete understanding has not yet been achieved. That is, along the lines of the discussion in the Preface and Chapter 1, we argue that much of design can be viewed as a cognitive process; thus, it can be modeled with increasing success. One of the major contributors to this success is our use of an AI-oriented approach in which we try to simulate the design process on a computer, an activity that forces us to articulate and externalize what we do when we design things.[♦]

An AI-oriented approach to investigating design processes offers the powerful advantage that the need for software implementation forces the forming of concrete models and thus helps eliminate vagueness, as in "then a miracle occurs" (Harris 2011).

2. *A successful representation – perhaps consisting of some ordering of multiple representations – can be found for both form and function*, and there exists a calculus for their interaction. Representing function is one of the most difficult problems in design representation, particularly so in our context of exploring

whether we can render something computable and so make it explicit.♦ (We must keep in mind that what we can do in our head may be distinct from what we can articulate to a computer because we apparently have adequate representations in our head – even if we are not always able to articulate them.) Relating function to form or shape is perhaps still more difficult, at least in the following sense. For an artifact of given form or structure, we can usually infer the purpose of the artifact. However, given an intended function that must be served, we cannot *automatically* deduce *from the function alone* what form or structure the artifact must have. For example, if we were given pairs of connected boards, we can examine them

> Representing function, and reasoning about it, is still considered to be "one of the most difficult problems," but there has been a lot of recent progress. Borgo et al. (2009) and Goel et al. (2009) provide a good current view, and two recent special issues of *AIEDAM: Artificial Intelligence for Engineering Design, Analysis and Manufacturing* (Stone and Chakrabarti 2005) and a chapter by Wood and Greer (2001) provide references to most of the additional key research in the field, including the continuing work on a "functional basis" (i.e., a standardized set of functional terms for products).

and deduce that the devices that connect them (e.g., nails, nuts and bolts, rivets, and screws) are fastening devices the purpose or function of which is to connect the individual members of each pair. However, if we were to start with a statement of purpose that we wish to connect two boards, there is no obvious link or inference that we can use to create a form or structure for a fastening device.

This does not mean that given a catalog of fastening devices, we cannot choose one to accomplish a given intention. In fact, we will see in Chapter 5 that among the attributes we can attach to device descriptions will be function and intent. But this is helpful only after we have already associated a purpose with a given device whose structure or form is known to us. Constructing a structure based on function alone is not, generally speaking, a process that is well enough understood that we can model it.

Note that we began this explanation with reference to a representation or to "some ordering of multiple representations." This is because we recognize that the same information can be cast in different forms, so there is no need to restrict ourselves to using a single representation.

3. Given a successful representation broad enough to span form and function, *the original statement of a design problem, and in particular its objectives and any applicable constraints, can be cast in terms of this representation.* This, in a sense, is a self-evident truth, for we should have to conclude that the representation is not especially robust – or useful – if we cannot recast our design statement, design objectives, and given constraints in terms of that representation.

4. *There exist problem-solving techniques that exploit this representation for the generation and enumeration of design alternatives.* This too is fairly obvious for, as we remarked in Section 1.1, there is not much point in looking for new representations unless we can effectively exploit them to do design more effectively. This also may distinguish our "engineering" approach from

a more "scientific" approach because we here are interested in the utility of new representations.

5. *The generated designs can be translated from the multiplicity of representations into a set of fabrication specifications.* Remember that the end point of a successful design is the production of a set of plans for making the designed artifact. This set of plans, the fabrication specifications, must have the properties of unambiguousness, completeness, and transparency; that is, it must stand on its own as thoroughly understandable (cf. Section 2.1).

6. *Criteria for design evaluation can be stated and applied in terms of either the representation used in the problem-solving phase of the design process or the formalisms used for the design and fabrication specifications.* Here, we are noting simply that we can assess and evaluate our design at different points in the design process; consequently, we may use different versions of our representation. It is worth adding, however, that we make no assumptions about whether the evaluation criteria are deterministic in nature or whether they involve uncertainty.♦

If uncertainty is a part of our evaluation criteria in a particular design problem, we should remember that it is often the case that the manufacture or use of a device points up deficiencies that were not anticipated in the original design. That is, successful designs often produce unanticipated secondary or tertiary effects that become ex post facto evaluation criteria (e.g., the automobile being in some sense judged a failure because of its contributions to air pollution).

Our definition of engineering design and its underlying implicit assumptions clearly rely heavily on the notion that some sort of representation, formalism, or language is inherently and unavoidably involved in every part of the design process. From the original communication of a design problem, through its mapping and solution, to its evaluation and fabrication, the designed artifact

> Designers have to face uncertainty throughout their design work. The designers may be uncertain about the priorities given to preferences, or even whether something is really a requirement; make deliberate assumptions not knowing whether those assumptions are "true enough" in this design situation; make selections from alternatives (e.g., a material) where that selection may not be overwhelmingly the best; be unable to know exact dimensions of parts due to manufacturing variations; and be unable to anticipate the operating conditions for the product, leading to uncertainties about expansion due to temperature, about wear patterns, and about environmental corrosive effects on joints and surfaces. Such uncertainties will "stack up," like tolerances, to produce a design that is riddled with uncertainty about its quality.
>
> Furthermore, designers may not take the time or make the effort to calculate, research, simulate, prototype, or infer everything they need to know: the result is uncertainty. There also may not be any theory, or adequate models, about manufacturing processes or material behavior, for example, which also produces uncertainty.
>
> A summary of representing uncertainty and reasoning under uncertainty can be found in Russell and Norvig (2010), and a general introduction to uncertainty management for engineering systems planning and design, such as Taguchi's Rough Design methods, is provided by de Neufville (2004).

must be described; it must be "talked about." Thus, *representation is the key issue*. It is not that problem solving and evaluation are less important; they are extremely important, but they too must be expressed and implemented at an appropriate level of abstraction. Thus, they are also inextricably bound up with concepts of representation.

Before leaving this chapter, we note that it is open to question as to whether there is a meaningful difference between specifications and constraints. Indeed, Simon has argued that the distinction is ephemeral.♦ The difference seems to hinge on whether we are referring to attributes of the designed artifact that we choose to optimize, in which case we speak of specifications, or whether we are working with attributes for which "satisficing" (see Chapter 6) is acceptable, in which case we speak of constraints. Because this distinction, meaningful or not, is maintained in most of the design literature, we maintain it in our own discussions.

> Distinctions often are made among needs, requirements, constraints, and preferences for design. The "needs" are what the customer wants, usually in incomplete terms that often mix functional, behavioral, and structural properties, all expressed at highly varying levels of abstraction. For software design, in particular, "requirements" are *testable* properties of the resulting design or its performance during use. "Constraints" are sometimes distinguished from requirements in that requirements specify what must be true, whereas constraints specify what must not be allowed to be or to become true. "Preferences" provide a basis for selecting a design from a set of designs, each of which satisfies both the requirements and the constraints.

2.3 Bibliographic Notes

Section 2.1: Cross (1989) presents a short discussion of design history and distinguishes design from a craft. Generate and test and other AI-based problem-solving methods are detailed in Dym and Levitt (1991a).♦ Palladio (1965) is the still-available translation of Palladio's works. Discussion of design in other fields includes architecture (Alexander 1964; Salvadori 1980; Stevens 1990), organizational decision making (Simon 1981; Winograd and Flores

> A modern treatment of AI is presented in Russell and Norvig (2010), and the art of concurrent engineering is described by Salomone (1995).

1986), and professional consultation (Schon 1983). Definitions of design are given by, among many others, Agogino (1988b), Dixon (1987), Jones (1970), Mostow (1985), Simon (1981), and Winograd and Flores (1986). Algorithms for aesthetics are presented in Stiny (1978). Two discussions of the collapse of the Hyatt Regency in Kansas City in 1981 are Petroski (1982) and Pfrang (1982). Levitt, Jin, and Dym (1991) describe both needs and architectures for concurrent engineering. Fenves (1988) and Rehak (1988) stressed the importance of fabrication specifications.

Section 2.2: The definition of design was first presented in Dym and Levitt (1991a) and was extended in Dym (1992b). The ephemeral nature of the difference between constraints and specifications was pointed out in Simon (1975).

3   Characterizing the Design Process

Having offered our own definition of engineering design, we now go on to study design as an activity – that is, the process of design. In this chapter, we review some of the established models of the design process. Then we go on to describe more recent articulations, some of which are rooted in the AI-based ideas mentioned earlier. Parts of the discussion may seem vague or abstract because we are trying to describe a complex process by breaking it down into smaller, more detailed pieces, but we are not going to produce a detailed cookbook that must be followed in order to complete a design. We are simply trying to picture, in words and diagrams, what is going on in our head when we are doing design.

3.1 Dissecting the Design Process

We start by looking back at the questions we asked (and others that we might have but did not) in our ladder-design exercise. In so doing, we find that we can *decompose* or break down the process into a sequence of steps by extracting and naming some of those steps. For example, when we ask, "How much weight should a safe ladder support?" and "For what purposes is the ladder to be used?," we are

> *clarifying the client's requirements.*

When we ask, "What is a safe ladder?" and "What is the allowable load on a step?," we are

> *identifying the environment* (or at least some aspects of the environment) in which our design would operate.

When we answer questions such as "What is the maximum stress in a step when it carries the design load?" and "How does the bending deflection of a loaded step vary with the material of which the step is made?," we are

> *analyzing* or *modeling* the behavior of the ladder.

When we ask, "Can this ladder be assembled?" and "Is the design economically feasible?," we are

> *identifying constraints*, including manufacturing, economic, marketing, and other constraints.

When we answer questions such as "Can the ladder carry the design load?" and "Is the ladder safe as it is actually designed?," we are

testing and evaluating the proposed design(s).

When we ask "Is there a more economical design?" and "Is there a more efficient design (e.g., less material)?," we are

refining and *optimizing* our design.

Finally, when we present the final fabrication specifications for the proposed design (and typically also the justification for those specifications), we are

documenting our completed design for the client.

We see quite clearly that the questions we are asking about the ladder design can be recognized as steps in a process in which we move from a fairly abstract statement of a design objective through increasing levels of detail until we can "build" a model of the ladder, perhaps optimize and refine some of its features, and then complete the process by documenting both the fabrication specifications and the justification for this particular design. In delineating the process this way, we are able to identify the specific tasks that need to be done to complete a design – and this, again, is the point of the exercise.

We should also note that as we work through these steps, we are constantly communicating with others about the ladder and its various features.♦ When we question the client about desired properties, for example, or the laboratory director about the evaluation tests, or the manufacturing engineer about the feasibility of making certain parts, we are interpreting aspects of the ladder design in terms of the languages and parameters that these experts use to do their own work. Inasmuch as the design process cannot proceed without these interpretations, we see once again that representation issues are central both to describing the object being designed and understanding the process by which the design is carried out. We look now at some descriptions of the design process.

> The focus of the formal study of engineering design has changed as research has progressed and matured, moving from restricted, often artificial "toy" examples of designing toward larger, more realistic examples. Studies of design have also moved from routine parametric design toward non-routine problems and design creativity, from single designers to design teams, from collocated activity to distributed activity, and from prescriptive theories to studies of designers in situ.
>
> Regarding teamwork and communication, Visser (2006) describes the need for "interdesigner compatible representations" such as sketches or gestures, and Fruchter and Maher (2007) provide a snapshot of the research on support for design teams. Aurisicchio et al. (2006) studied the information requests of engineering designers because such queries are one of the reasons why designers are "constantly communicating with others." Visser (2006) emphasizes the fact that individual designers are still designing even when they are in collocated teams, although they undertake additional activities that facilitate cooperation. Hence, research on individual design activity remains very important.

It should be stressed that referring to *the* design process should be done with great care. This entire chapter is about *characterizing* design. As a result, we present abstractions of typical processes. In real life, one or more agents generate designs, guided or limited by constraints, preferences, evaluation knowledge, and a wide variety of knowledge and data. These limits derive from various sources, such as the knowledge, skill, and experience of the designer(s); available tools and methods; externally imposed needs and requirements; and limitations imposed by the physics. Hence, the context for a particular design process can be infinitely varied, leading to wide variations in the actual process followed by different designers – even for the same requirements! The designer's knowledge and experience affect the "routineness" of a problem (Brown 1995). Gero and Kannengiesser (2007) describe how the developing design solution, as well as the members of the design team, acts to "situate" the process.

One of the simplest models of design considers only transformations between types of representations. For example, Gero's (1990) model of design as a process connects the required functions (F), the behaviors expected given F (B_e), the resulting design structure (S), the behaviors that S can produce (B_s), and the final design description (D). It defines the processes of *formulation* (F to B_e), *synthesis* (B_e to S), *analysis* (S to B_s), *evaluation* (comparing B_e to B_s), *reformulation* (S to B_e), and *production* of the design description (S to D). The model does not prescribe the order in which these processes occur and does not suggest the kinds of reasoning that might be used for each type of process. Subsequent developments include a more complex model, as well as the use of this model to annotate protocols collected from designers (i.e., a detailed record of their activity) to support visualization and analysis of their activity.

3.2 Describing the Design Process

One of the simplest and most straightforward models of the design process consists of three stages.[♦] In the first stage, *generation*, the designer proposes various concepts that are generated, by means unspecified. In the second stage, *evaluation*, the design is tested against the design goals, constraints, and criteria that have been set forth by the client and the designers. In the third stage, *communication*, the design is communicated to the manufacturers or fabricators. Although this model has the virtue of simplicity, it actually tells us relatively little about what goes on. It is sufficiently abstract that it does cover the examples we have already discussed, but not nearly in enough detail to advise us on how to proceed with a design. We cite one obvious question, *How* do we generate designs? In fact, answering just this question could be said to be the *raison d'être* for this book and all the other books and articles ever written about design!

One of the most widely cited models of the design process is displayed in Figure 3.1. In this model, the circles usually represent some form of description of the evolving design, although they sometimes represent a stage (e.g., as in the statement of needs that begins the process).[♦] The rectangles indicate some design activity (e.g., analyzing a problem or doing a detailed design). As in our structural engineering example, we begin with a stated *need*, which first is elaborated through a process of questioning the client and gathering relevant data, until we arrive at a clear statement of the problem. Then we enter the *conceptual design* stage, in which we look for different concepts (or *schemes*) that can be used to solve the stated design problem. In a bridge design, for example, our different schemes might be represented by different bridge types, say a

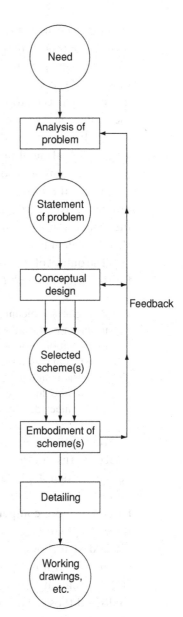

Figure 3.1. A pictorial view of the design process (French, 1992).

classic suspension bridge, a cable-stayed bridge, or an arch. Our assessment of these three schemes would depend on some of the high-level attributes of our design goal, including the anticipated span length, financing, type of traffic, and aesthetic values of the client.

The conceptual design stage is the most open-ended part of the design process: discussion is relatively abstract in this stage and the focus is on high-level trade-offs between possibly conflicting goals. Thus, it is here that we are most likely to see negotiation among participants who have very diverse interests in order to resolve performance goals and other trade-offs. For example, at this stage in a bridge design, we might see structural engineers pitted against investors, traffic analysts arguing with the authority that wants to build the bridge, and various specialist designers haggling over priorities. At this stage, we probably focus more strongly on the function of the artifact than we do on its form – notwithstanding the bridge schemes just cited – except to the extent that all the participants in the design process see the artifact as a variation on a known theme or a mutation of a familiar artifact or design. In conceptual design, too, we cannot predict how subsystems will interact, which subgoals might conflict, which options may have to be ruled out because of local conditions, and so on, because we have not yet refined our design very much. Again, for the bridge design, the nature of the gap being spanned (e.g., Are we bridging a river or crossing over one or more roadways?) could result in variations in the span length, which in turn could result in different configurations of the secondary structural members.

The output of the conceptual-design stage is a set of possible concepts or schemes for the design. French defined a *scheme* as:

By a *scheme* is meant an outline solution to a design problem, carried to a point where the means of performing each major function has been fixed, as have the spatial and structural relationships of the principal components. A scheme should have been sufficiently worked out in detail for it to be possible to supply approximate costs, weights and overall dimensions, and the feasibility should have been assured as far as circumstances allow. A scheme should be relatively explicit about special features or components but need not go into much detail over established practice.

In the structural design problem in Section 1.3, our choice of a steel frame and a steel roof truss for the mill building would be such a scheme. For the ladder, we might have a wooden stepladder as our scheme. Note that although we have identified only one scheme in each of these two examples, the output of the conceptual stage could be two or more competing schemes. In fact, some would argue that the output of the conceptual stage *should* be two or more schemes because early attachment to a single design choice is viewed as a mistake. This tendency is so well known among designers that it has produced the aphorism: "Don't marry your first design idea." In any event, at this point in the process, we probably do not have sufficient data to discard those schemes that will later be viewed as "extra" because they did not measure up for one reason or another. For example, we may still be entertaining the notion that the stepladder be made of either wood or aluminum or perhaps some more exotic material.

The next phase of the design process is called, variously, the *embodiment of schemes* and *preliminary design*. This is the part of the process in which the conceptual proposals are first "fleshed out"; that is, we hang the meat of some preliminary choices upon the abstract bones of the conceptual design, or, in French's connotation, we embody or endow the conceptual design with its most important attributes. In this phase, we begin to select and size the major subsystems based on lower-level concerns that include the performance specifications and the operating requirements. For the ladder, for example, we would begin to size the side rails and the steps, and perhaps decide on the way that the steps are to be fastened to the side rails. In the mill-building example, we would lay out the roof truss in greater detail, including locating the roof purlins, estimating the size of the truss-joint connections, and so on. In preparing such a preliminary design, we might well use various estimates or back-of-the-envelope calculations and algorithms as well as rules about size, efficiency, and so on. This phase of the design process makes extensive use of rules of thumb that reflect the designer's experience. And, in this phase of the design, we make a final choice from among our candidate schemes.

The penultimate stage of French's model is *detailing*, or as it is usually called in the United States, *detailed design*. Here, our concern shifts to refining the choices made in the preliminary design; our early choices are articulated in much greater detail, typically down to specific part types and dimensions. This phase of design is quite procedural in nature, and the procedures themselves are well known to us. Much of the relevant knowledge is expressed in very specific rules as well as in formulas, handbooks, algorithms, databases, and catalogs. As a result, detailed design has benefited greatly from attempts to encode accessible databases within CADD tools. This stage has also become rather decentralized; that is, it is left almost completely to component specialists as the design moves much closer to being assembled from a library of standard pieces.

The final stage illustrated in Figure 3.1 corresponds to the last stage in our dissection of the ladder-design process. We regard the production of "working drawings, etc." as equivalent to producing fabrication specifications for the assembly of the design and a design justification for the client. However, we should note that although this model of the process is more detailed than the simple one we discussed at the beginning of this section, we are not much closer to knowing how to do a design. Both of the process outlines we have given are descriptive; that is, they describe what is being done rather than detailing what ought to be done.

3.3 Prescriptions for the Design Process

We now present two prescriptive models of the design process. These models differ from the two descriptions given in Section 3.2 in that they *prescribe* a set of tasks that must be completed to generate a satisfactory design. In the two descriptive models of the design process that we have just seen, we can recognize within them three generic tasks that are repeatedly performed, usually in an iterative fashion, within each phase of the design process. These three tasks are:

Synthesis is the task of assembling a set of primitive design elements or partial designs into one or more configurations that clearly and obviously satisfy a few key objectives and constraints. Synthesis is often considered as the task most emblematic of the design process.

Analysis is the task of performing those calculations (or analyses) needed to assess the behavior of the current synthesis – or embodiment or preliminary design. For example, we calculate the bending stress and deflection of a loaded step of the ladder to see how the step behaves under a given set of loads. There is another argument to be made that analysis is a task also done early in the design cycle because analytical thinking is required to clarify the client's requirements and lay out the specifications against which the design will be made and evaluated.

We are doing the *evaluation* task when we compare our analyses of the attributes and behavior of the current design to the stated design specifications and constraints to see if this synthesis is acceptable.

These task descriptions do provide a general statement about *what* we are actually doing when we design an artifact, but they are still fairly general. There is a "mixed" model of the design process wherein the prescriptive tasks are incorporated within descriptions of phases of the design process, as follows:

1. The *analytical* phase of the design process involves two activities:

    In the *programming* activity, we identify those issues that are crucial in the design, in a manner very much akin to our earlier suggestions (cf. Section 3.1) about how important it is to clarify the client's objectives. Here, too, a plan is proposed for completing the design process, perhaps through the construction of a schedule of milestones and deliverables for the design project.

    The second activity we perform in the analytical phase is *data collection*, in which we collect and collate as much relevant data as possible and perhaps organize it into a design database that will be the data source for the duration of the design project.

2. The *creative* phase of the design process involves three activities:

    Analysis is the first activity in the creative phase.♦ As with our previous description of this task, it calls for logical thinking and modeling to help us decompose the design problem into subproblems; complete the translation of the client's requirements into design specifications; review and revise, if necessary, our project schedule; and prepare a budget for the design process.

    > We must be careful, however, not to confuse the word *creative* with *creativity*. Here, in this very abstract prescriptive model, we use the term only to refer to the stage in which the actual design is generated (i.e., *created*).

    The *synthesis* task as outlined in this model is quite similar to our previous description of this task in that it calls for the preparation of skeleton designs that are, essentially, equivalent to conceptual designs or French's schemes.

    Development of the candidate designs is the final creative task. It includes both the evaluation of alternative conceptual designs and the preparation of preliminary designs or prototypes.

3. The *executive* phase of the design process involves only one activity:

    This final activity, *communication*, is akin to the task of documenting the final manufacturing or fabrication specifications.

This mixed model depicts the design process as a "creative sandwich." In the first or analytical stage, we require organization, analysis, and data collection.

In the final or executive stage, we create fabrication specifications in an objective and orderly fashion. However, the middle stage, the creative phase, is a sort of puzzle. On the one hand, according to the description, the creative task depends on the application of analysis and logic. On the other hand, this stage also requires the exercise of subjective judgment and personal involvement in developing schemes or concepts. In truth, it is this part of the design process that is the most difficult to model or represent. Even in the clearest books on design, this phase is discussed only in very general descriptive terms; the design methods described are, in fact, only design aids that are intended to help us be creative and to cast as wide a net as possible. However, as we have noted before, the boundaries between what we can and cannot now model are not static, so we should not be deterred by the apparent thickness of the creative filling in this design sandwich.

Designers and scholars in Germany have put forward a more detailed set of prescriptions that reflect an attempt to systematize and formalize the design process. Two such models are displayed in Figures 3.2 and 3.3. The first reflects four stages, each of which consists of several tasks and produces specific outputs. In brief:

1. The *clarification* phase has as its output a *design specification*.

   This goal is achieved by *clarifying the task* presented by the client and *elaborating a specification* in sufficient detail so as to define a specific target toward which we can aim our design effort and against which we can, eventually, measure our success.

2. The *conceptual design* phase has as its output a *concept*.

   This goal is achieved by *identifying the most crucial or essential problems*; *establishing a function structure* – that is, a framework within which the artifact will perform its primary function, including a decomposition of the primary function into subfunctions that will be performed by subsystems or individual components; *formulating a solution procedure* that can be (successfully) applied to the design problem; *preparing concepts* or skeleton designs or schemes; and *evaluating candidate schemes* against the relevant criteria, including both economic and technical metrics.

3. The *embodiment design* phase of the design process actually has two stages, the first of which has as its goal the production of a *preliminary layout* and the second of which has as its output a *definitive layout*.

   The preliminary layout is obtained by *refining the conceptual designs, evaluating and ranking them* against the design specifications, and *choosing the best* as the preliminary design.

   The definitive layout, the final output of the embodiment phase, is obtained by *optimizing the preliminary design* and by *preparing preliminary parts lists and fabrication specifications*.

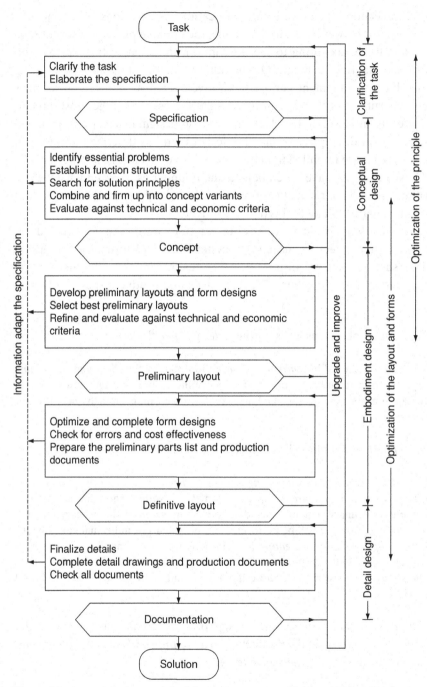

Figure 3.2. A prescriptive model of the design process (Pahl and Beitz, 1984).

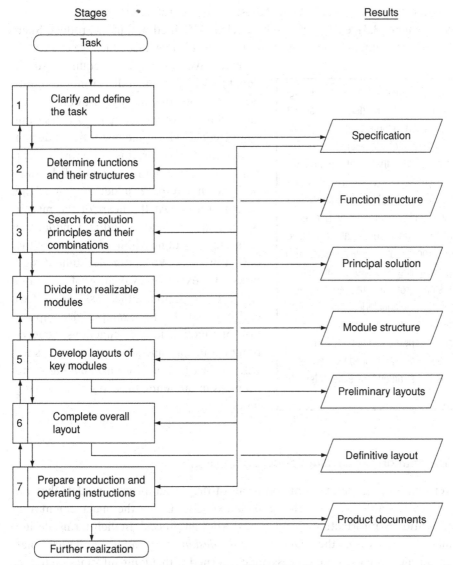

Figure 3.3. The design process as outlined in *VDI–2221: Systematic Approach to the Design of Technical Systems and Products* (Cross, 1989).

4. The *detailed design* phase, the final stage of the design process, has as its output the *documentation* of the designed artifact.

> This goal is achieved by *checking our results* and *documenting our design* in final manufacturing or fabrication specifications.

This model, the most elaborate of those presented thus far, is still not the most complex. The German technical society of professional engineers, the Verein Deutscher Ingenieure (VDI), has articulated a number of guidelines aimed at making design more systematic. One of these guidelines refines the process into seven stages (cf. Figure 3.3), each of which has a clearly defined output or product. Such a systematic refinement is perhaps useful as a checklist against which we can check that we have done all the "required" steps, and it may have some utility from the standpoint of protecting an organization against liability for a product design or for assuring conformance with some institutional design approach.♦ However, it is not entirely clear that this or any other more detailed elaboration adds much to our understanding of the design process. At the heart of the matter is our ability to understand and model the separate tasks done within each phase of the design process. We can model some design tasks – for example, evaluation of a scheme and documentation. Other tasks are not so easily modeled – for example, decomposition of function into subfunctions, generation of concepts, and so on. However, as we will see in Chapter 6, we are making some progress in modeling these problem-solving tasks.

We show yet another prescriptive model in Figure 3.4. It is based on a set of models of five stages in a design process. The inputs to any one stage are dealt with by a set of stage-specific design tasks that produce outputs that form the inputs to the next stage (Dym et al. 2009). These depictions are not intended to imply a strict linear process that is entirely without iteration, as Dym et al. remind their student readers. Despite allowing backward loops in the process, prescriptive models tend to suggest that designing is a sequential process that moves in steps from abstract to concrete. However, Visser (2006) concludes that "empirical cognitive design studies show that *systematic* implementation of stepwise refinement is rare in practice," despite the fact that top-down, breadth-first refinement is the strategy that experts tend to favor.

3.4 Information Processing Models of Design

One very interesting feature common to all of the design process models that we have surveyed – although only rarely made explicit – is that the management and transformation of information is central to the design process. In their terms, design information derives from the *adaptation of the design specification* – and recall that even the earliest stages of the process are concerned with adding information to the client's statement for the purpose of clarifying the client's objectives and substantiating a design specification. In this light, it is particularly interesting to describe two information-processing models of design.

The first such model we discuss is called the task-episode accumulation or TEA model. The principal components of this model are a design state and a set of design operators situated within a *design environment*, a "map" of which is shown in Figure 3.5. The *design state* incorporates all the information about a design as it evolves. The most important parts of the design state are (1) proposals for achieving design goals, and (2) constraints that the design must meet (e.g., specifications) or within which the design must operate. The *design operators* are primitive information processes that modify the design state as various design tasks

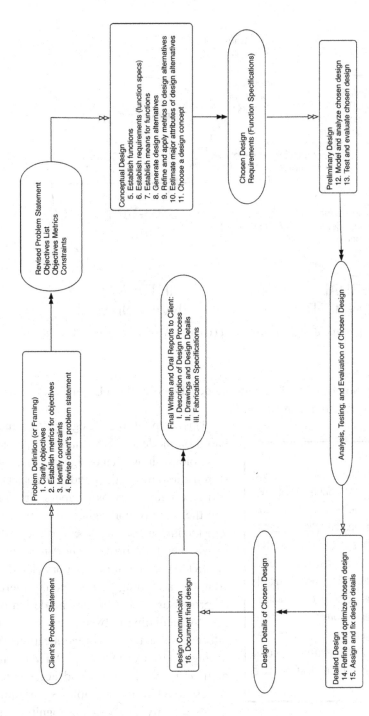

Figure 3.4. Elaboration of the design process that characterizes stages in terms of their inputs, tasks performed, and outputs (Dym et al., 2009).

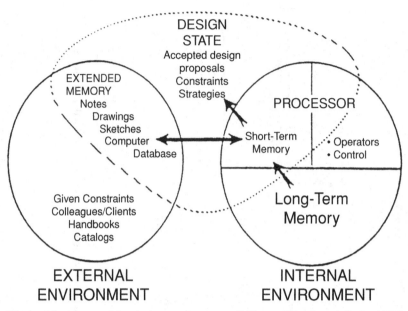

Figure 3.5. A map of the design environment (Ullman, Wood, and Craig, 1990).

are completed. Some of the 10 operators developed in the TEA model are select, create, simulate, calculate, and so on.♦

This map of the design environment, based on models of information-processing psychology, shows that design information is stored internally, in the mind of the designer, and externally, in books, papers, drawings, and the mind of colleagues. We can subdivide the internal storage locations according to whether they provide short-term memory (STM) or long-term memory (LTM). In addition, according to this model, there is a "processor" that controls the design process by mediating between and among the internal and external storage locations. We distinguish between STM and LTM because of their respective sizes and access times. Access to STM, for example, is very quick, but its capacity is very small. Cognitive stud-

Although the TEA model lists 10 operators, Brown (1992) lists more than 30 different roles for knowledge-based reasoning during designing, including abstraction, classification, criticism, evaluation, patching, planning, and selection. This difference may be due in part to the fact that TEA is intended to be an information-processing model, whereas Brown's list is driven by types of knowledge and their use. It is actually quite difficult to choose a "right" level at which to describe designing sufficiently well enough to build useful computational models. For example, describing design as "analysis, synthesis, and evaluation" is too abstract.

ies seem to show that only seven *chunks* or meaningful units of information can be stored in STM at any instant, and that the chunks of experts are usually larger than those of novices. Conversely, LTM has an essentially unbounded capacity but the access time is fairly long, on the order of 2 to 10 seconds per chunk.

Furthermore, access to LTM is achieved by some triggering mechanisms or operators that act on information brought into STM (cf. Figure 3.5). Inasmuch as *all* design problem-solving information (or, for that matter, any cognitive processing information) must pass through the STM, we could infer that STM could prove to be a significant bottleneck for human problem solving – a fact that has obvious implications for designing computer-based designer's assistants (cf. Preface and Section 1.2).

We do not intend to go into further detail about the TEA model, in part because its authors view it as emerging and still under development, and in part because some of its details underlie a design taxonomy that we discuss in Section 4.3. In that taxonomy, we articulate the design state, at various levels of abstraction, and the roles of some of the operators in greater detail. It is interesting to note, however, that this model has emerged from a conscious attempt to extend cognitive psychology models into the design domain.

In the second information-processing model, design tasks are proposed and analyzed in terms of the information they use. We note two basic sets of processes in this approach. The first set of processes *propose* design choices. The second set is a collection of *auxiliary* processes that *provide information* in support of generating or testing a design choice or commitment. As we delineate these two sets of processes, we introduce more of the terminology common to AI-based research in design problem solving (see Chapter 6):

1. There are four basic information processes that *propose design commitments* – that is, processes that generate design choices or make commitments about specific aspects of a design. They are:

   A basic and often-used approach to design is the *decomposition* or reduction of a problem into a set of smaller problems. We thus intend to solve the overall design problem by solving subproblems for which we have previously compiled or known solutions or that require searches of much smaller solution spaces.

   We can also view design as a process of *design planning* – that is, of formulating a plan of steps that need to be taken to produce the fabrication specifications for a designed artifact. The steps in the plan could be set out as goals to be achieved or in terms of the parts or components of the artifact.

   In the process of *design modification*, we explicitly recognize the existence of previous designs that we can modify to meet our current design needs. We do need criteria for choosing designs that are close to the current problem, and it would be helpful to know why an old design fails to meet current goals and what must be modified so that we can adapt an old design.◆

   > Case-based reasoning (CBR) systems (Maher et al. 1995; Maher and Pu 1997) can index cases (e.g., design plans) by the goals that they satisfy and by the problems that they avoid. Such CBR systems can repair failed plans and build up a set of features that can be used to predict problems (Hammond 1989).

> Constraint processing (Dechter 2003) is now a well-developed area as well as a natural fit for parametric design (O'Sullivan 2006) and configuration problems (Felfernig et al. 2011) that can be handled by assignment (Wielinga and Schreiber 1997).

Constraint processing or manipulation can be a useful tool for design when the structure of the artifact is known and the design process can be reduced to selecting values for variables and parameters.♦ This kind of information processing is often done algorithmically because the problem has sufficient structure to make numerical-modeling approaches such as optimization useful.

2. There are four basic types of information processing that *provide auxiliary information* – that is, information in support of generating or testing a design choice. These processes are:

> The task of translating high-level goals and constraints into more detailed versions appropriate to the subproblems is generally very difficult to execute: effectively, goals and constraints must also be decomposed and made more specific. This is particularly difficult, for example, with constraints on parameters that are accumulative. For example, if *x* is composed of *a* and *b*, and *x* must be less than 10 inches, then it is difficult to be more precise about either *a* or *b*. Such conjunctive goals are very difficult to deal with while designing, or repairing failing designs, due to the inherent dependencies.

The process of *generating goals and constraints for subproblems* is very important as we translate high-level goals and constraints into more detailed versions appropriate to the subproblems of the original decomposition.♦

Recomposition is the process of assembling solutions to subproblems into a coherent solution to the overall problem. Much of the information required here is about how the pieces fit together.

Design verification is that part of the design cycle in which we test our solution to verify that it meets all goals and constraints.

Design criticism is information used to analyze failures, identify their causes, and suggest modifications to fix them.

In closing this brief overview of information-processing models of design, it is worth noting that the *auxiliary* information processes are fairly typical problem-solving skills that can be invoked equally well to solve analysis problems; that is, they are not unique to design. The processes involved in *generating* design choices, conversely, are relatively specialized toward design, where the idea is to create either an artifact or the specifications for an artifact. It is thus not surprising that the modeling of these design processes has become a fertile branch of research for the AI community, in large part driven by the recognition that algorithmic processes – suited as they are to aspects of evaluation, verification, optimization, and documentation – are simply inadequate in terms of providing the information needed to generate designs.

3.5 Design Methods in the Design Process

The descriptions of the design process given here are neither unique nor exhaustive. There is significant overlap among these process descriptions, the taxonomies of design we present in Chapter 4, and the problem-solving techniques we discuss in Chapter 6. This merely reflects the complexity of the design concerns that come into play in analyzing the design process.

We have also noted that these process descriptions often fail us because they do not tell us *how* to go about the business of generating or creating designs. Although we can be rather specific – to the point of being algorithmic – about some aspects of the design process, we have a difficult time with design generation. We address this point in greater detail in Chapter 6 but, having pointed out so often the limitations of these process descriptions, it seems worthwhile to introduce here some more traditional, inductive design methods, casting them in terms of the AI-based problem-solving methods that are discussed in much greater depth in Chapter 6. As a context for this discussion, we use the process description paralleling that shown in Figure 3.1.

One method used to *clarify* the original project statement is the construction of an *objectives tree*. Here, we make a hierarchical list – which actually branches out into a tree-like structure – in which all of the objectives that the design must serve are ordered by degree of specificity. Then, starting at the highest level of abstraction, we have our top-level design goal, which is the *end* the design must serve. As we move down the list to more detailed and specific objectives, we find that we are generating the *means* by which the design will perform its desired function. Thus, this method is related to a general problem-solving technique called *means-ends analysis*.♦

> Means-ends analysis (MEA) is a technique for reducing a problem by mapping a *difference* to an action or actions that should help reduce that difference. The "difference" is usually between a current state and some desired state (e.g., a goal or subgoal). MEA is thought of as recursive; that is, reduced differences may need to be further reduced.

Conceptual design can be thought of in terms of searching a large space of possible designs. Because the space is likely to be both large and complex, we do not want to look for solutions in an ad hoc fashion because we may miss good solutions or, at best, take a very long time to find them. One method often proposed for generating conceptual designs or schemes is *functional decomposition*.♦ The idea behind this method is to decompose the primary function that the design is intended to perform into subfunctions, which in turn are decomposed until some level is reached in which the design of subsystems or components that perform these subfunctions is

> The standard functional terms in the functional basis (Hirtz et al. 2002) are organized into a hierarchy of different levels, thus providing support for refinement and perhaps for decomposition. Decomposition typically requires reasoning, but CBR can be used. Decompositions can result in subfunctions that are either independent of order or dependent on some behavior that is part of another subfunction (Erden et al. 2008).

relatively clear. *Decomposition*, or "divide and conquer," is a way to reduce a large problem into a set of smaller – and presumably easier – subproblems.

Of course, in this process we have to keep in mind the interactions between the subproblems and to monitor their individual solutions to ensure that they do not violate the assumptions or constraints of the other, complementary subproblems.

Another general approach, particularly for conceptual design, is to follow a strategy of *least commitment* – that is, to make as few commitments as possible to any particular configuration because the data available are perhaps too abstract or very uncertain at this point in the design process. Or, it may be that the data are simply not available so early in the process. Least commitment is less a specific method than a strategy or a (good) habit of thought. It militates against making decisions before there is a reason to make them. The dangers of making premature commitments are perhaps obvious; that is, that we could either limit in a suboptimal way the range of designs that might solve a particular problem or that we could become wedded to a concept that develops unsuitably and from which we must then withdraw – perhaps after a significant amount of wasted effort. As a strategy, least commitment is of particular importance in early stages of design (e.g., conceptual design), where the consequences of any one design decision are likely to be propagated far down the line.♦ In later stages (e.g., detailed design), we are very interested in making it easy to test and fix design decisions (see the following discussion).

> It is often stated that as much as 80% of the life-cycle cost of a product is determined by design choices made at the conceptual design stage. However, with about 80% of the design effort still remaining at that stage, we cannot estimate costs easily or accurately unless we can draw analogies to similar designs. However, every commitment we make can be used to generate hints about the consequences for life-cycle cost.

In preliminary design, we are concerned with generating candidate solutions and then with either testing them to ensure conformance with design objectives and applicable constraints or evaluating them against metrics – for example, cost – that support a choice among otherwise adequate designs. One common reasoning strategy is *generate and test*, in which we devise a generating strategy that automatically generates large numbers of potential designs that are then tested against the relevant metrics and constraints. With this strategy, however, we must guard against a *combinatorial explosion* of the space of candidate designs in which we are simply overwhelmed by the number of possible designs that emerge as a consequence of different combinations of the design variables.

Often, having generated a solution that fails to meet a specified test, we want to redirect our search for a solution so as to fix the failure. In particular, in detailed or final design, we can often identify the decision(s) that precipitated a failure. In such cases, we can formally apply a problem-solving approach called *backtracking* to remedy the problem by undoing the decision(s) that caused the failure. This requires that we explicitly articulate the *dependencies* or links beween design decisions and the values of the artifact attributes that result from those decisions.♦ Such design methods often use explicit statements of design *constraints* along with mechanisms to *post* or propagate them to appropriate points in the assembly of final design attributes.♦

We have focused this short discussion of design methods on the *thought processes* or *cognitive tasks* that are done during the design process. This brings us closer to explaining *how* we actually go about designing artifacts, and it sets the stage for articulating a taxonomy of design tasks, as we do next.

> VT, an expert elevator configuration design system (Marcus et al. 1987), maintains dependencies between decisions. When constraints fail, VT uses the dependency record to guide knowledge-based backtracking to undo the decision that caused the failure. This approach does not systematically undo the last decision made each time because it is a waste of time if that decision was not the actual cause of the problem. The dependency record can also be used to discover how many subsequent decisions such a change might affect. Other AI-based systems use implicit or explicit dependencies in a similar way.

3.6 Bibliographic Notes

Section 3.2: Models of the design process were proposed by Cross (1989) and French (1985, 1992). French (1992) introduced the term *schemes* for conceptual design. The importance of generating multiple schemes in conceptual design is pointed out in Pahl and Beitz (1984) and VDI (1987). The pithy aphorism on this point was brought to our attention by Antonsson (1993).

Section 3.3: The tasks of design are outlined in Asimow (1962), Dym and Levitt (1991a), and Jones (1981). The "mixed" model of design was originated in Archer (1984) and reviewed in Cross (1989). More extensive design prescriptions representing a German approach to systematizing design are found in Pahl and Beitz (1984) and in VDI (1987).

Section 3.4: The TEA model of design was developed by Ullman, Dietterich, and Stauffer (1988). It is based on models of

> Constraint posting and propagation are included in the MOLGEN system (Stefik 1981). Constraints can be determined and then attached (i.e., posted) to a future decision (e.g., a variable) so that they can act as evidence to help decide what the decision result (e.g., the variable's value) ought to be. In a "least commitment" system, posting constraints allows decisions to be made only when enough evidence has been gathered such that the result is clear. MOLGEN reasons about plans for experiments. It can take a constraint on an output of a plan action and reason about what must have been constrained in the action's input; that is, the constraint is propagated back "through" the action. The ultimate type of constraint is a value (e.g., $X = 10$) because it acts as a constraint on the values of other variables, with propagation possible via equations (e.g., $X + Y > 20$).

information-processing psychology developed in Newell and Simon (1972) and Stauffer (1987). Another information-processing model of design was developed by Brown and Chandrasekaran (1989) and is described in reasonable detail in Dym and Levitt (1991a).

Section 3.5: Traditional methods for design (e.g., objectives trees) are reviewed in Cross (1989) and French (1992). Means–ends analysis as a problem-solving style is propounded in Newell and Simon (1963, 1972). Functional decomposition is described in Cross (1989) and Ullman (1992b), and "divide and conquer" as an AI technique is detailed in Rich (1983).

4 Taxonomies of Engineering Design

We have now defined, at least tentatively, what we mean by design, and we have described several views of the process of design. In so doing, we have identified some of the ways that problem-solving strategies could be employed in the design process. Now we turn to the task of trying to outline an organizational structure or taxonomy for design. According to the dictionary, a *taxonomy* is the result of the "study of the general principles of scientific classification" and a *classification*, at least in the natural sciences, is the "orderly classification of plants and animals according to their presumed natural relationships." Similarly here, a taxonomy of design might (1) allow us to classify design problems according to certain characteristics; and (2) facilitate the organization of the knowledge, representation, and reasoning schemes that would be useful in modeling different kinds of design.

Are such taxonomies important? Why? One viewpoint is that a scientific theory of design cannot be developed without such a taxonomy. Another is that such taxonomies allow us to compare design methods and design tools – especially those reflecting the newer computer-aided technologies the ideas of which are reflected in this discussion. Our own viewpoint is stated somewhat differently in that we would stress any increase in our ability to understand and model the thought processes involved in design as being the best reason for developing such taxonomies. As we work toward this objective, we will simultaneously increase our abilities to compare various design tools and to develop and explore new design methods. We note also that, as with our characterizations of the design process, various taxonomies and classifications have been proposed, and there is certainly some overlap of ideas among them. After we review these various taxonomies, in an order roughly corresponding to their chronological development, we will try to analyze in a principled way what we have learned from so describing the thought processes of design.

4.1 Routine versus Creative Design

We know that experience gained from previous attempts to design the same or similar artifacts allows an engineer to solve a design problem more easily and efficiently. Through such experience, we acquire knowledge that helps us relate some

The distinction "routine versus creative" has been drawn by many researchers (e.g., Brown and Chandrasekaran 1989) over the years, but it now appears to be confusing at best, if not simply wrong. This is because identifying a design as creative renders a judgment on that design, relative to personal or group norms (Boden 1994). Thus, as strange as it may seem, being creative probably depends as much on who is making the judgment as on the design process itself: *any* design activity *may* produce a design that may be judged to be creative. More formally, creative design has been defined as requiring a "transformation" of the search space (i.e., changing the reasoning rules), as described by Boden (1994) and Gero (1990).

Highly nonroutine design is characterized by not knowing how to proceed and by a lack of immediately available appropriate knowledge. This situation typically requires new activity that requires problem solving and much less reliance on existing methods. This has been called "innovative design," but the term has been defined in various ways. The distinction "routine versus innovative" is better than "routine versus creative," although highly innovative design does tend to lead to creativity. More recently, "innovation" has come to mean the *implementation* of creative concepts.

of the design requirements to parts or subsystems of the artifact that can achieve those requirements. Similarly, past experience helps us learn how to assemble and organize independent parts to achieve a desired overall behavior. Such cases – in which we have a lot of the knowledge about parts, components, systems, and their functions – involve relatively routine design in which the problem-solving task is largely a process of matching requirements to previous attempts at meeting the same set of requirements.[♦] Although we might have to repeat this process several times in an iterative fashion to achieve an acceptable design, it is still a relatively simple task.

Oftentimes, however, the design goal is a device that is quite different from previous designs, or the number of alternatives that must be considered is very large, or the desired artifact is completely new. In such cases, the design activity is more complicated: the space of possible design solutions is then both large and complex. In some cases, the space of possible designs may not be easily or well defined. Therefore, any attempt to generate and test all possible designs would require a prohibitively large effort, so we need powerful ways to search through a large, complex space of possible designs. We would expect in such situations that the knowledge obtained either from experience or from a deeper understanding of the domain plays an essential role in guiding our search for plausible design alternatives. Obviously, we may also use such knowledge to modify a design if some of our design choices turn out to be undesirable.

The foregoing discussion suggests that we can characterize the type of design we may be facing in accordance with an assessment of the knowledge we have about the design domain and of the strategy we will use to solve the particular design problem at hand. The knowledge assessment depends on how complete is the knowledge we use to generate designs – especially with regard to the degree of specificity we can attach to the form, goals, and constraints governing our design problem – and how complete is our store of additional or *auxiliary* knowledge that we require to test possible designs. The second significant factor for assessing design complexity has to do with how difficult it is to control the problem-solving process we use to search the space of possible solutions. Thus, we will describe first a taxonomy that is based on asking whether a design is *routine* or whether it is *creative*. This classification is more or less equivalent to assessing how difficult it is to generate candidate designs. Specifically, then, three classes of design are identified in this taxonomy:

Class 1 Design. Also known as *creative* design. Creative design is rare in that it usually leads to completely new products or inventions. It is typically characterized by a lack of both domain knowledge and problem-solving strategy knowledge. Here, the goals are vague, and we are short both of ways to effectively decompose problems and of designs for the subproblems. This kind of design requires considerable problem solving even in its auxiliary processes. It is original, rare, and almost certainly not susceptible to encapsulation with current representation technologies – probably because we do not understand the origins or form of true creativity.

Class 2 Design. The distinguishing mark of such design, sometimes called *variant* design, is that we know a lot about the design domain; that is, we understand the sources of our design knowledge but we lack a complete understanding of how that knowledge should be applied. Thus, although we may be able to successfully decompose a design task into a number of subproblems of component design, it is the modification or replacement of the individual components that makes the design difficult to complete. The continuous revisions in automobile design that we have seen in the last several years provide a classic example of this. Automobile basics are still very much the same, but the individual components are certainly quite different – and much more complex – than their counterparts of even a decade ago. Thus, although automobiles still have engines, transmission systems, and wheels, the subsystems that operate and connect them are much more sophisticated and complicated than they were only a few years ago.

More recently, the intermediate class of designs – between creative design and routine design – has been termed *innovative design*, although the emphasis in this definition varies somewhat. One view is as just expressed; that is, in innovative design, we lack a clear-cut problem-solving strategy. A second view is that innovative design perforce hinges on the application of reasoning from first principles – that is, from the fundamental physical equations and concepts used by engineers. The two views can be said to converge if we assume that the domain knowledge we know consists of the fundamental physical knowledge, whereas the problem-solving strategy we lack is the knowledge of how, where, and when we should apply this fundamental knowledge.

Class 3 Design. In *routine* design, we typically know in advance everything we need to know to complete a design. That is, we can identify the specific design or domain knowledge we need to complete the design (or, we know the sources of that design knowledge) and we know how to apply that knowledge. Thus, we typically would have effective ways to decompose design problems, well-understood and efficient (compiled) plans for designing components, and fairly complete information about the possible causes of design failures that can be appropriately applied during the design process. Still, routine-design

A three-class model is too abstract to accurately describe all activity that occurs during design. However, if we make the description less abstract by adding more classes and details, then we are less likely to match every example of design: there are just too many individual variations due to requirements, constraints, preferences, resources, and knowledge available. (This, by the way, is one reason that algorithmic and decision-based models of design are limited in their ability to portray the entire scope of the design activity.) Thus, it seems that there is always some sort of difficulty when describing design!

Note that the three classes in the Brown and Chandrasekaran (1989) model are not distinct: they depend on two variables, domain knowledge and problem-solving strategy knowledge, each of which can vary independently and continuously. In addition, it is important to recognize that any characterization of design activity applies only to a portion of the whole because each subproblem itself (and its subproblems in turn) might be Class 1, 2, or 3; that is, each subproblem might be more or less routine. It is better to characterize the situation at any single point in time during design (Brown 1996) because the situation can vary as the designer switches from subproblem to subproblem, recognizes what knowledge is appropriate, reasons out a plan to tackle some design subproblem, retraces steps after failing, and so on.

Judgments about the routineness or difficulty of a design problem can be made by an individual designer, design team, or design community. Similar judgments are also made about creativity: they can be made before, during, or after a design project, and about the designed artifact or the design process.

problems often require a significant amount of design knowledge because there may be complex interactions among subgoals and among components, as a consequence of which we must anticipate complexities both in selecting and ordering design plans and in undoing steps already taken to undo failures (in the jargon of AI, we call this *backtracking* or *retracting design commitments*). Thus, even in the simplest class of routine design, there is more than ample scope for the intelligent deployment of design knowledge.

This taxonomy is interesting and suggestive but, unfortunately, its simplicity is limiting – the three classes of design do not appear to carry enough distinctive detail to be helpful for characterizing the wide variety of activies that fall under the rubric of design.♦ Furthermore, the identification of a design problem as routine or variant or creative depends on the experience of the problem solver who is doing the classifying. Routineness is thus a relative measure. What is routine for one designer may not be for another. Furthermore, given that designers appear to learn a great deal from experience, a problem that seemed difficult two years ago may now be seen as routine by any given designer. Routineness is thus an "individual's standard," measured in "the brain of the beholder."

In addition, there is also a "community standard."♦ The professional design community may consider an engineering design problem routine – meaning that there is an expectation that the problem will be routine for all practicing members of the community. The expectation also may derive in part from the current state of design education if, for

example, the specific domain knowledge and problem-solving strategy for solving a particular design problem or class of problems are generally taught in design courses.

Applying the community standard is itself not a trivial task, although it is probably easier to see as design problems become less routine. Clearly, the community standard for a particular design problem is represented by the pool of existing design solutions. Thus, a design would be deemed "innovative" relative to that pool of existing designs. However, it is perfectly possible for a design to be innovative relative to an individual's standard but not so relative to the community standard. Thus, design problems are innovative in context, not in and of themselves, so we should be cautious about labeling innovative design.

Similar reservations could be expressed about the term *variant*. In some of the literature, it has been used to describe Class 2 designs as we have just done. In process planning, the term *variant* has been used to describe the process of choosing a plan from among a set of standard plans. Similarly, in design, the term could be taken to mean producing designs by varying prototypical designs. This describes how a design is produced but makes no judgment about its quality, originality, or routineness. If we were to think of variant designs being produced as a result of changing parameter values only, without changes in function features, form features, or topology, we would probably not classify these problems as innovative.

Finally, and as noted previously, routine design is still often sufficiently complex – and difficult to model computationally – that most KBES design applications have reflected attempts to provide advice for carrying out Class 3 (routine) design tasks. Some of these applications are described in Chapters 5 and 6, but first we move on to describe other design taxonomies.

4.2 A Taxonomy of Mechanical-Design Problems

As noted earlier, one argument for developing design taxonomies is to further the opportunities for generalizing design knowledge into a coherent formal theory and methodology of mechanical design. This argument contends that an identifiable design process exists only in a fairly abstract way, as we have seen in Chapter 3, and that there are many different design processes that are carried out operationally, depending on the particular design problem, the designers involved, and the environment in which the design is done.[*] The mechanical-design taxonomy proposed by Dixon and his colleagues is an attempt to classify design problems, as a starting point for further discussion aimed at developing a formal theory of design.

Chandrasekaran's (1990) analysis of design tasks considers that every task can be carried out by multiple methods, each of which might include other tasks. Thus, design proceeds recursively through tasks and methods, with the selection of methods being context dependent. His analysis proposes that tasks can be of the type *propose*, *verify*, *critique*, or *modify*. So, in a highly familiar situation, a *propose* subtask might be handled using case-based reasoning (i.e., applying a previous, similar design case, albeit with minor alterations). In a less familiar situation, a *verify* subtask might be done by simulation – for example, by verifying that a particular behavior occurs. This recursive model clearly explains the myriad possibilities for design activity.

The "state of knowledge" referred to here concerns what is currently known about the final design. The state can be viewed as moving from knowing very abstract aspects to knowing very concrete aspects, such as particular parameter values. Typically, the knowledge state also can be seen as moving from desirable effects on the user's environment, via function, to behavior and then to structure. This allows design to be defined as progressing from a more abstract state to a less abstract state. This progress can be viewed as being orthogonal to the degree of routineness: some state changes may be easy and some difficult.

At the highest level of abstraction, a design problem is characterized here by the specification of two knowledge types: an *initial state of knowledge* and a *final desired state of knowledge*.♦ These two states of knowledge are each in turn characterized as being one of six mutually exclusive *knowledge types*, with the types chosen from a list of seven knowledge types. Once the initial and final states are specified, we will identify problem types as a function of the differences between the knowledge types of the two states. The seven mutually exclusive knowledge types used to define the two problem states are:

Perceived need. This is the condition or need that provides the motivation for designing something. Whereas perceived needs are often expressions of social or economic needs, engineering design is here limited to more refined, less abstract expressions of *functional* need. And, it should be noted, not all design problems begin at this abstract level; that is, many are stated at more refined or detailed levels. For our stepladder example (cf. Chapters 1 and 3), the perceived need could be that we want to provide a means to allow people to obtain access to heights exceeding their own, whereas the engineering-design version of the perceived need would be a statement of the need to provide the ability to support a given weight at a certain specified height.

In the taxonomy that we discuss in the next section, it is also noted that specifications are design requirements or goals that are based on the perceived need. Such design specifications may be based on functional performance or they may relate to other constraints, such as spatial constraints, manufacturing restrictions, cost, or applicable codes and standards.

Function. This is the most detailed statement of the perceived need that can be made without reference to physical principles, form, or embodiment (see the following discussion) or specific artifact types. A function indicates what must be done without specifying how it is to be achieved. A *functional requirement* is a translation of the function into a detailed, quantitative, operational statement. For the stepladder, the functional requirements could state that among other things, each step should support a person weighing 350 lubes and the weight of the ladder should not exceed 15 lubes.

Physical phenomena. This is a statement about the underlying physical principles that will be used to design the artifact, although this statement is done without reference to how these physical concepts will be displayed in the actual object. In the German literature, physical phenomena are called the *working principles*. Thus, the stresses in a step that supports forces on the step might be restricted to be significant only in directions transverse to the step thickness or supported force.

Dixon et al. make an interesting point here, which is that the physical phenomena just specified for the step are often subsumed in the statement that the step should respond as a bent beam or plate, in which case we are permitting a phenomenon's *name* to be substituted for the phenomenon itself. This abstraction, although at times very convenient, may in fact inhibit thinking by short-circuiting the range of options being considered. Although not especially realistic for a ladder step, the statement given would allow us to design the step as a beam or as a truss, whereas we would ordinarily perhaps have just required the step to be a beam.

Embodiment (or **concept**). Concepts or embodiments represent a generalized form or shape based on the physical phenomena being developed to achieve a function. Thus, a ladder step can be seen as a beam, although the beam's cross section is not yet specified.

Artifact type. An artifact type represents a concept refined further, wherein the specific attribute types of the concept – although not their values – are detailed. Thus, the beam for the step might be specified as a thin plate or as an I-beam or perhaps as something else again, as long as the artifact type behaves like a beam.

Artifact instance. Here, specific values are given for the artifact type, thus creating an *instance* of the artifact; in current jargon, we are *instantiating* the artifact. Thus, we could specify that the step be made of wood and have dimensions of 30 inches × 8 inches × 0.50 inch.

Feasibility. This is simply an assessment of whether a particular aspect of a design is deemed feasible.

Then, the following two design-problem states are defined:

The *initial state of knowledge* is one (and only one) of:

> *perceived need*
> *function*
> *physical phenomenon*
> *embodiment*
> *artifact type*
> *artifact instance*

The *final state of knowledge* is one (and only one) of:

> *function*
> *physical phenomenon*
> *embodiment*
> *artifact type*
> *artifact instance*
> *feasibility*

Problem Type	Initial State	Final State
Functional	*perceived need*	*function*
Phenomenological	*function*	*phenomena*
Embodiment	*phenomena*	*embodiment*
Attribute	*embodiment*	*artifact type*
Parametric	*artifact type*	*artifact instance*

Figure 4.1. A taxonomy of types of major design problems (Dixon et al., 1988).

Now, we need to remember that the complete specification of a knowledge state could, in fact, incorporate more than one of the six types relevant to it, depending on whether it is an initial or a final state. However, when we define particular problem types, it is on the basis of a transformation from a single initial knowledge state to a single final knowledge state. With this in mind, Dixon and his colleagues propose a taxonomy of five major design problem types (Figure 4.1). Problem types are identified according to the difference – which is normally a refinement – between discrete initial and final states of knowledge.

This taxonomy of major problem types does not name all the possible combinations of design problems because, as the authors note, "there are too many possible combinations to conceive unique, memorable names for all of them." However, the definitions of knowledge states can be used to identify some of the more familiar terms used to describe design problems. For example, for preliminary and conceptual design and design feasibility,

Preliminary Design:	*function*	*artifact type*
Conceptual Design:	*function*	*embodiment (concept)*
X-Feasibility Study:	*X-initial state*	*feasibility*

It is easiest to think of examples of routine design occurring in situations where simple choices or basic calculations are being made – for example, when a value for a parameter (i.e., a length) is determined. Thus, there is a strong association between parametric design and routine design. However, design is more likely to be routine when sequencing and decomposition knowledge exists in addition to decision-making knowledge. This can occur *anywhere* in the abstract-to-concrete spectrum.

It also follows, then, that a complex design process can be decomposed into a series of the basic problem types. We will see in Sections 4.3 and 4.7 that such decompositions can be used to identify design tasks identified in other taxonomies and that these decompositions may not be unique. We could, for example, identify routine design as the following combination:♦

Routine Design:	*embodiment*	*artifact type*
	and	*artifact type* *artifact instance*

To provide some concreteness to this discussion, we show in Figure 4.2 one example of a knowledge state that might be relevant to the design of wooden

Need	To assemble structures with wooden parts
Function	To fasten adjoining wooden parts together
Phenomenon	Friction
Embodiment	Metal part inserted into adjoining wooden parts
Artifact type	Nail
Artifact instance	Six-penny galvanized nail

Figure 4.2. An example of a knowledge state definition (Dixon et al., 1988).

stepladders. Here, we could identify as routine design the process of choosing a concept or embodiment (e.g., a metal part inserted into adjoining wooden parts), choosing a particular artifact type (e.g., a nail), and then finally instantiating the nail by identifying a particular size (e.g., a six-penny nail). Clearly, other choices could be made in the knowledge-state definitions that would lead to other design choices. For example, the phenomenon chosen could be bonding, in which case the embodiment would be a bonding agent, the artifact type could be a wood glue, and the artifact instance would be a particular type or brand of wood glue.

There are subclassifications of this top-level design taxonomy; for example, we could refine an artifact type as the specification of a *physical type*, an *assessment type*, and a *complexity* or *coupling*. We do not need such detail at this point, so we close by noting that the taxonomy presented in this section focuses on classifying aspects of the design problem, as distinct from the design process that entails both the people involved (e.g., the designers) and the design environment in which they work and the design tools at their disposal.

4.3 A More General Mechanical-Design Taxonomy

We now turn to a mechanical design taxonomy proposed by Ullman that extends the taxonomy just discussed in several ways. The particular focus of this work is to enable not only the formalization of design methods and theory but also the comparison of design tools, especially the more recent computer-aided design environments that are beginning to appear as commercial products. Furthermore, this taxonomy is set in the context of the TEA design-process model discussed in Section 3.4. It is thus relevant both to our continuing focus on the design thought process and as an elaboration of the TEA model.

Ullman explicitly includes the design environment and the design process in his taxonomy, arguing that a complete taxonomy must encompass more than the defined artifact. There are three top-level components: the *environment*, including the characteristics of and the constraints on those doing the design; the *problem* being solved, as defined by its initial and final states; and the *process* by which the initial problem state is changed to its final state. Each of these components is further refined as shown in the table displayed in Figure 4.3. The slots in the right column are filled in as various options are identified.

If we now look at the problem part of the taxonomy, we see that it extends the previous one in a particularly interesting way; that is, the initial and final states are

Top-Level	First-Level Refinement		Values
Environment	Participants		1. _____
	Characteristics		2. _____
	Resources		3. _____
Problem	Initial State	Refinement Level	4. _____
		Representation	5. _____
	Final State	Refinement Level	6. _____
		Representation	7. _____
	Satisfaction Criteria		8. _____
Process	Plan		9. _____
	Processing Action		10. _____
	Effect		11. _____
	Failure Action		12. _____

Figure 4.3. A form for a generalized mechanical design taxonomy (Ullman, 1992a).

detailed in terms of the representation in which they are expressed as well as their own refinement levels. The latter are chosen from the same list of six possible initial knowledge states used in the previous taxonomy and applied here to both initial and final knowledge states; that is,

> The *initial state of knowledge* and the *final state of knowledge* are each characterized by one (and only one) of:
>
> *perceived need*
> *function*
> *physical phenomenon*
> *embodiment*
> *artifact type*
> *artifact instance*
>
> The representations indicate the "language" in which that knowledge is stated. Four such *representation languages* are identified:
>
> *textual*
> *numerical*
> *graphical*
> *physical*

The first three of these languages are self-explanatory; the fourth could refer to a physical model or perhaps to a set of equations or some other mathematical realization that is used to model the underlying physical phenomena. Thus, each refinement of an initial or final state will have an accompanying "value" of representation language in the appropriate slot (i.e., Slots 5 and 7 in Figure 4.3). We say more about the representation languages of design in Chapter 5.

Another addition to the problem definition is the identification of the criterion that must be satisfied to declare a design complete. We like to think of designs being *optimal* in some sense; that is, they are the best designs possible when judged by some stated criteria. However, the state of the art of optimal design is such that it normally can be applied only when there is a clear-cut mathematical model of the design objectives. In reality, however, we are quite pleased if we can *satisfice* – that is, if we can find a design solution that is satisfactory when judged by reasonable, often nonquantitative, measures that are usually less restrictive than those applied in formal optimization. Furthermore, there is empirical evidence that many designers work toward just such ends.

We now turn to the process part of this taxonomy, one that is identified as "still emerging" as research in the area unfolds. Four components are identified as essential for characterizing a design: the *plan*, the *processing plan*, the *effect*, and a *failure action*. The refinements of these categories are displayed in Figure 4.4 and are detailed as follows:

> Planning during design might be done in a variety of ways, from mere plan execution at one end of the spectrum, via plan instantiation (e.g., using a parameterized plan), to plan construction from sub-plans, to full knowledge-based planning. An introduction to AI–based planning is found in Hendler et al. (1990), and Ghallab et al. (2004) provide a complete survey.

Plans. Four different kinds of design planning are identified.[♦] They are design by *fixed plan*, in which the design proceeds in a fully delineated process, as a "cookbook" approach; *selection of a plan from a list of plans*, in which instance designers choose a plan from a set known to them from previous experience;

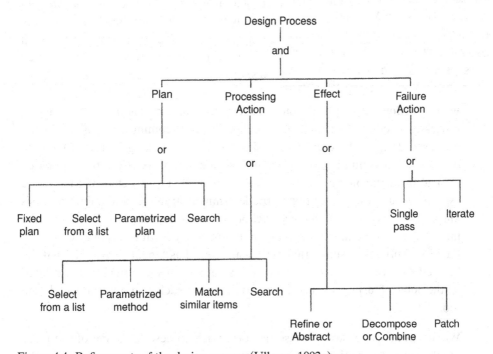

Figure 4.4. Refinements of the design process (Ullman, 1992a).

parameterized plan, in which case a skeletal plan outlining the major steps is used (with details filled in by the designer) or a single plan is applied with a control strategy based on values of specified design parameters; and *search*, in which instance the techniques prominently used in AI are used explicitly to search a space of design solutions. These search methods can be *weak* or *strong*, depending on the degree to which domain knowledge is used to guide the search.

Processing actions. Some kind of action must be taken to execute the design plans just identified. These processing actions are also four in number and mirror closely the design plans; that is *selection from a list*, wherein potential solutions can be chosen from a list of previously known designs; *parameterized methods*, in which the design problems can be represented in terms of a set of mathematical equations whose solutions depend on various design parameters that are specified in the description of the problem's initial state; *matching of similar items* is a kind of search that depends strongly on reasoning by analogy; and (again) *search*.

Effects. Design processes have three identifiable effects on designs: *refinement*, in which the design final state is more detailed than is its initial state; *decomposition*, in which the design problem is broken into smaller subproblems for which the strength of the coupling between subproblems is a major issue; and *patching*, in which the design is modified in some way that does not reflect a true refinement, such as where a longer screw is substituted when a shorter one will not do.♦

It is also possible for designers to move in a direction opposite to *refinement* to produce an *abstraction* of the current design situation – that is, a higher-level, less detailed view of the design. More abstract views are used to recognize specific methods or general strategies that might be useful or even to recognize previous designs in the same general category.

Failure actions. The fourth and final category in the classification of design processes is concerned with the handling of shortcomings or failures in the proposed design. There are two basic approaches to fixing failures. One is to stop the process and proceed to apply some external fix. The other approach is to go through some iterative process to fix the failure by changing the failed aspects through the application of predetermined strategies and failure-analysis knowledge. (We see in this last dichotomy the role, albeit implicit, that the intended use of the taxonomy plays in its construction. In terms of design tools, the second process assumes that the "design tool" has "an internal logic" that can be used to iterate to a more satisfactory solution. This notion clearly is related to the idea of using the taxonomy to compare such computer-aided design tools.)

We are now in a position to use this taxonomy to describe some of the more commonly used descriptions of the design process. In doing so, we follow the

Top-Level	First-Level Refinement		Values
Problem	Initial State	Refinement Level	Perceived need or function
		Representation	Textual
	Final State	Refinement Level	Concept
		Representation	Textual or graphics
	Satisfaction Criteria		_____

Figure 4.5. The completed taxonomy form for *conceptual design* (Ullman, 1992a).

classification form shown in Figure 4.3. However, unlike Ullman's presentation, we (1) delete all of the environment part of the form, consistent with our interest in the design thought process; and (2) delineate the process part of the form only in those cases where it is filled out.

Thus, *conceptual design* is defined simply in terms of the problem alone without reference to either the design environment or the processes applied (Figure 4.5). We see that the refinement level and the representation of both the initial and final states are, in fact, fairly abstract.

Selection design is that process by which a designer selects one or more parts or components from a list (e.g., a catalog), all of whose members have the requisite attributes and satisfy the appropriate criteria (Figure 4.6). This particular design characterization is interesting also because it shows up in a task-oriented taxonomy that we describe in Section 4.4.

In addition to presenting completed forms for other design processes (e.g., layout and detail design, parametric design, a task-episode accumulation model, and the VDI–2221 design process described in Section 3.3), Ullman presents completed forms for the "whole model" of routine design (Figure 4.7) as well as for those portions of a routine design that would be performed by human specialists (Figure 4.8) and by an automated "specialist" – that is, a computer design tool that could run automatically without human intervention.

In filling out the taxonomy forms for Class 3 design (cf. Section 4.1), all three of the top-level classes are included because some of the key distinctions involve both the design environment and the design process. This analysis assumes that we begin with knowledge about how to decompose a complex object into components

Top-Level	First-Level Refinement		Values
Problem	Initial State	Refinement Level	Artifact instance
			Parts list
		Representation	Textual or graphics
	Final State	Refinement Level	Artifact instance, part
		Representation	Textual or graphics
	Satisfaction Criteria		Satisficing OK
Process	Plan		Fixed plan
	Processing Action		Selection
	Effect		Refine
	Failure Action		Iterate

Figure 4.6. The completed taxonomy form for *selection design* (Ullman, 1992a).

Top-Level	First-Level Refinement		Values
Environment	Participants		Computer-assisted design; single user
	Characteristics		Unknown
	Resources		_____
Problem	Initial State	Refinement Level	Concept
		Representation	Textual
	Final State	Refinement Level	Artifact instance, assembly
		Representation	Textual
	Satisfaction Criteria		Satisficing OK
Process	Plan		Mixed
	Processing Action		Mixed
	Effect		Refine
	Failure Action		Iterate

Figure 4.7. The completed taxonomy form for *routine design*, "the whole model" (Ullman, 1992a).

and that our knowledge of the components is also relatively complete. However, the overall taxonomy (Figure 4.7) serves only to disguise all the knowledge about decomposing both artifact and process, and it adds very little to our ability to describe what is happening – and for the same reasons that we found the routine–variant–creative taxonomy itself to be limited. The detail is what makes it interesting, but the taxonomy for the whole routine-design problem disguises that detail.

However, if we refine the routine problem into what the human designer does in decomposing the problem into properly posed subproblems, each of which is done by an automated "specialist," then we find a great deal of information in the "human decomposition" of routine design (Figure 4.8). It is the human designer who takes the initial concept (in Ullman's example, a table) and refines it into a set of artifact types (say, a top and one or more legs) that can be designed by the specialists. Jumping from a concept or function to an assembly of artifacts or artifact types is

Top-Level	First-Level Refinement		Values
Environment	Participants		Individual designer
	Characteristics		Unknown
	Resources		_____
Problem	Initial State	Refinement Level	Concept
		Representation	Textual
	Final State	Refinement Level	Artifact type, assembly
		Representation	Textual
	Satisfaction Criteria		Satisficing OK
Process	Plan		Search
	Processing Action		Matching similar items
	Effect		Refine and decompose; weakly coupled
	Failure Action		Iterate

Figure 4.8. The completed taxonomy form for *routine design*, "human decomposition" (Ullman, 1992a).

not uncommon in mechanical design because we have a wealth of experience with many mechanical devices. However, this refinement is often a very difficult process, which is one of the reasons that detailed taxonomies may be useful for exploring and understanding design.

4.4 A Task-Level Taxonomy for Selection Design

This short but very pragmatic taxonomy of design was originally presented as part of a discussion of roles for KBES technology in improving design. This taxonomy includes the selection of configurations – as well as components – and it includes as a major element the characterization of the search space in which a design solution eventually will be found. In each of its four categories, we apply a selection process to design components and configurations. However, note how the characterizations given here differ from those that appear in Figure 4.6.

> **Component selection.** Here, we choose a component from among a set of available components the number of which defines the search space.

> **Component parameter design.** Here, we choose values for parameters for a component in order to meet stated requirements, such as in V-belt drive design. The size of the search space is characterized by the number of parameters, which is typically small.

> **Configuration selection.**[*] Here, we organize or assemble a known set of components into a specified architecture. The size of the search space, which may be quite big, is defined by the number of feasible combinations of the components. One example of configuration selection is the configuring of VAX computers for assembly.

> **Configuration design.** Here, we organize a known set of components into an architecture that has not been defined in advance. This is the most complicated task in this taxonomy because both the components and the architecture in which they are placed may have to be designed rather than simply selected. The number of parameters could be very large and their values may vary continuously or discretely, such as in the design of paper-handling subsystems in copiers.

> Wielinga and Schreiber (1997) describe many tasks that might be termed "configuring" – that is, producing a configuration. Although standard definitions of the configuration task require a predefined set of components, those components may be fully or partially specified. By varying the specificity of the components provided, the strength of the prescription of the desired configuration, and the nature of the requirements and constraints, it is possible to describe a variety of possible types of configuration design. There are eight major types, including *verification* (i.e., configuration checking), *assignment* (e.g., assigning people to rooms), *parametric design*, and *full configuration design*.

This pragmatic classification of selection design incorporates elements that are not captured in the taxonomy given in Figure 4.6. This is in part because Ullman's selection design is about selecting components; thus, it can emcompass only the first two elements of this taxonomy. The motivation for characterizing the search space as well as the kind of search is that it facilitates the comparison of computer-based design tools (one of Ullman's interests!). However, it is also slightly different because it suggests a different refinement of the processing actions (or problem-solving strategies).

4.5 Selection Design Refined

We now present a refinement of the process of selection design, adapted from a taxonomy of *methods for solving arrangement problems*.[♦] We present this abstraction in increasing order of specificity to parallel the creative–routine classification outlined in Section 4.1 (and we will see some overlap with that taxonomy).

> Brown (1998) suggests another view of configuring: *selecting* (choosing components) plus *relating* (establishing abstract relationships) plus *arranging* (establishing specific relationships) plus *evaluating* (testing the compatibility of components and verifying that goals have been satisfied).

We return to it in Chapter 6 because it will be helpful in identifying specific problem-solving methods within each category. We will also see that in the descriptions of some of its constituent elements, this breakdown incorporates problem-solving strategies in a very explicit way.

Creative design. This is the process of generating designs for truly unique products – that is, products that are not assembled from libraries of elementary components in relatively standard ways. Again, such design is still largely the province of human designers.

Assembling unique solutions from elementary components. This category refers to design problems the solutions of which are sufficiently distinct from one another that prototypes cannot be used economically. Solutions are typically assembled from elementary components to achieve a synthesis that satisfies all goals and constraints at some level. Inasmuch as the elementary components invariably interact with one another in significant ways (e.g., geometrically and functionally), there are numerous constraints to be satisfied among the components. Design by assembly requires knowledge of all elementary components, their attributes, their behavior, and their constraints with respect to other components.

Hierarchical generation, testing, elimination, and evaluation of solutions. Here, we generate all possible solutions at a relatively high level of abstraction, eliminate most of them by applying heuristics, generate all solutions at the next level of abstraction for those that survived the first elimination, and continue on recursively. In the final step, the surviving solutions are ranked using a set of evaluation criteria. This approach depends on our ability to generate and meaningfully test solutions and, as we noted in Chapter 3, can be viewed as a structured form of trial and error.

Prototype selection and refinement. Here, we include "semicustom" design in which a library of prototypical solutions serves as the starting point for solving a new design problem. If we have a good candidate for a prototypical solution, we can test it and modify those of its attributes that fail to meet constraints. This works well for the design of relatively standard products. We note, too, the argument that prototypes are the very stuff of design.

Pure selection. Here, we satisfy any imposed constraints by selecting an arrangement from a set of known alternatives. We thus require knowledge of all of the alternatives and their attributes. Pure selection is used to select standard components or subsystems in the detailed design stage of semicustom products.

4.6 Knowledge-Level Analysis of Design

We mentioned that one of the potential uses for design taxonomies is the perception that they could provide useful metrics with which to evaluate and compare KBES design applications. More than a decade ago, the abstraction of *knowledge-level analysis* was proposed for identifying the behavior of a KBES separately from the (symbolic) representation used to implement that behavior.[♦] This abstraction has been restated in other forms; for example, the domain knowledge used in a system is made up of the "components of expertise." Regardless of the particular statement of the abstraction, the aim is to identify a way of abstracting what a design system *does*, so that its domain-independent problem-solving methods can be identified apart from the domain knowledge used to solve design problems.

Analyses of the problem-solving performance of knowledge-based systems appear to stem from the identification of the domain-independent, problem-solving method of *heuristic classification* as underlying the work of one of the earliest successes in expert systems, the Mycin system, which is used by doctors to support the diagnosis of infectious diseases.[♦] Heuristic classification is built around the processes of abstraction, matching, and refining. In particular, heuristic classification includes the ability to *abstract* data, *match* symptom and diagnostic data in order to apply heuristics ("rules") provided by the domain experts, and obtain a solution – a diagnosis – by progressively *refining* both symptomatic and diagnostic data.

> Smithers' (1998) knowledge-level model provides a way to describe design in terms of the knowledge used. Types of knowledge include needs and desires, requirements statements, solution statements, requirements-formation knowledge, problem-solving knowledge, and problem-revision knowledge. Although this knowledge-level model *describes* different types of design, such as innovative design, it does not *prescribe* and does not explicitly include any control knowledge; rather, it merely indicates how knowledge might flow.

> Clancey's (1985) model of heuristic classification was part of a wave of research that looked for models of generic, domain-independent reasoning. Jackson's (1999) excellent – although now sadly out of print – text on expert systems describes some of that history. Heuristic classification can be part of design activity. For example, it can be used to classify a situation in order to select an appropriate method, plan, or previous design case. Klein (1991) uses heuristic classification to determine the type of conflict between agents in a multi-agent design system. Once the type has been determined, a mapping is made to a method to resolve that type of conflict: the more precise the classification, the more detailed the resolution strategy that can be suggested.

Chandrasekaran's generic tasks (GTs) were originally seen as building blocks from which practical
knowledge-based systems could be built (Chandrasekaran and Johnson 1993). In addition to
heuristic classification, the list of building blocks
included "hierarchical design by plan selection
and refinement." However, careful analysis reveals
that even the "hierarchical design" building block
included ingredient types of reasoning, such as
basic synthesis, criticism, decomposition, evaluation, and selection. This helps confirm the notion that
tasks and methods are recursive for anything other
than the most routine and often-repeated tasks.

It also can be argued that the symbolic-
representation techniques applied in extant
KBESs do, in fact, provide a straightjacket
because although they represent what is possible computationally, they might not be sufficiently flexible to describe the tasks that
designers actually perform. Chandrasekaran
proposes that higher-level, more generic
primitives better model how designers actually solve design problems. These *generic
tasks* can be combined into a generic
framework for design, called the *Propose–
Critique–Modify* family of design methods.[♦]
These methods consist of four generic tasks:
design, *verify*, *critique*, and *modify*, and they
can be performed in a variety of ways. The four generic tasks are similar to the processing actions described in Ullman's taxonomy, and they will be seen to be similar
to some mechanisms that are described next.

In a parallel context, Balkany, Birmingham, and Tommelein analyzed several configuration-design KBESs with the aim of identifying a structure for
comparing what such systems do. Their work focuses on parsing (from existing system descriptions) the problem-solving approaches used and the mechanisms devised to support the design problem solving captured in these systems.
Whereas our own motivation for examining taxonomies is, as has been repeatedly stated, to delineate the thought processes used in design, their analysis does
provide ideas that may be useful for refining some aspects of the taxonomies
already presented. The decomposition of design problem-solving knowledge here
includes:

Domain knowledge that defines an area of expertise.

Mechanisms or procedures that operate procedurally on well-defined inputs to
produce equally well-defined outputs.

Control knowledge, which is the complete collection of knowledge that is used
to properly sequence the mechanisms.

Problem-solving methods that operate as higher-level procedures that organize
the application of the mechanisms.

Tasks that represent specific applications of a problem-solving method.

Again, it is argued that the domain and the problem-solving method applied
are distinct and independent; that is, a problem-solving method could be applied in
several different domains. In this context, it is the last four of the previous categories
that are likely of greatest interest in refining, in Ullman's terms, the processing part
of a taxonomy. Conversely, problem solvers are strongly coupled to tasks and can

be applied effectively only to a single task, so that task analysis must clearly be a significant component of a process taxonomy.

Of particular interest in this decomposition are the mechanisms, which are articulated in terms of six functional groupings. Note that the names given to the mechanisms are expressed in a style derived from the symbolic programming languages in which such knowledge-based systems are written, but they are evocative of what the methods do. We are interested in them because they represent possible refinements of the design-processing actions that we have seen previously.

Select–design–extension mechanisms. These mechanisms apply two different schemes to extend the current state of a design. Preconditions are used to evaluate whether an extension to a current design can be made. Rankings are used to select by formula the best of several possible extensions, assuming that a single-number metric is available to order the respective degree of desirability. The availability of such rankings is, of course, suggestive of the use of formal optimization schemes.

Make–design–extension mechanisms. Here, formulas or algorithms are used to actually make the specific design extensions recommended by an extension mechanism. The extensions are made, in turn, by calculating values of design variables and then propagating them as needed. This is reminiscent of the parameterized design process outlined as part of Ullman's taxonomy.

Detect–constraint–violation mechanisms. Here, design extensions are evaluated against relevant constraints to ensure that they have not been violated.

Select-fix mechanisms are used to select a repair or fix of a design extension that fails when tested, as just outlined. In some KBES applications, the user of the system is notified that a failure has occurred and requests intervention to provide a fix. Other systems provide a menu of choices, including prompting the user, applying a precompiled fix (i.e., an efficient fix already in place), or generating a repair automatically. This set of mechanisms could be seen, then, as an elaboration of the category of failure actions identified in Ullman's taxonomy.

Make-fix mechanisms actually perform the selected fixes or repairs.♦

The selected fixes, or repairs, can be applied one at a time (*disjunctive*) or, if necessary, more than one at a time (*conjunctive*). Sometimes a fix can cause another constraint to fail and, in the worst case, fixing that violation can cause the first constraint to fail again (*antagonistic*). Klein (1991) points out that although the need for fixes might be "compiled out" of a system at build time (which is very difficult to do), many computational design systems use a "knowledge-poor" approach requiring that a predetermined set of fixes be used at each particular failure point. Such a set can be preordered by the amount of work each fix might entail or by the amount of damage they might do to the partial design already formed (Brown 1985; Marcus et al. 1987). In some situations, such as constraint failure, fixes can be reasoned out (e.g., if the test $A > B$ fails, then suggest an increase in A, a decrease in B, or both). In general, redesigning may degrade to designing, so experiential knowledge (e.g., design cases and heuristics) is vital to maintain efficiency.

Test-if-done mechanisms are applied to verify that a design, or part of a design, has been satisfactorily completed in terms of the goals set out at the appropriate level of abstraction – that is, at the appropriate level of component, subsystem, or overall design.

The decomposition of design knowledge and the specification of mechanisms just outlined have been used not only to compare some well-known KBESs for configuration design; they also have been used as the basis for building a domain-independent design environment. As indicated previously, these particular mechanisms may be useful for refining design-processing actions, although we need to keep in mind that they have been identified explicitly only for configuration design. Thus, their usefulness for generally explicating design processes may be limited. However, it is interesting to note that a similar set of mechanisms is used in the development of a system (CPD2-Soar) to aid in the design of chemical-distillation processes. Although intended for a different purpose and embedded within a very different architecture, the CPD2-Soar system uses operators to define and navigate its problem space. Acting much as the previous mechanisms, some of CPD2-Soar's operators are:

Get-feed operators. These operators interact with the user to define the feed stream that will be split by the distillation process.

Order-component operators rank the components of a stream in descending order according to their volatility.

Make-splits operators generate all the possible sharp splits that the system can apply to the feed stream.

Update-stream operators are used to compute the mole fractions of a stream's components, normalize their volatilities with respect to the heaviest component, and compute the total flow rate of the stream.

Although extrapolating from somewhat strained analogies is always hazardous, it does appear that the idea of mechanisms identified in this taxonomy could be extended and applied to other domains. We have more to say about this in Section 4.8.

Finally, on the topic of knowledge-level analysis, we make explicit once more a theme developed in Chapter 3. All of the ideas we have described can be viewed in the larger context of trying to identify what we know, so that we can then think about how best to express that knowledge usefully in one or another circumstance. It is interesting that one aspect of this is the argument that truly integrated engineering computational environments cannot be built without our having an explicit understanding (e.g., a taxonomy!) of the kinds of knowledge we have and the options from which we can choose when representing a given knowledge type. One application of explicitly different knowledge types has been proposed for the domain of structural design, and the utility of an explicit taxonomy of design knowledge has been raised.

4.7 Analysis of Design Tasks

The flavor and orientation of the last analysis of design we review have much in common with Brown's three-level classification (cf. Section 4.1) and with the knowledge-level analysis of Balkany, Birmingham, and Tommelein (cf. Section 4.6). However, we also point out that the design-task analysis we now describe is placed squarely by its authors in the context of software engineering in which AI-based approaches are viewed as a methodology for doing useful software engineering for design. Tong and Sriram propose that *design tasks* can be classified along several dimensions, including:

Available methods and knowledge. This dimension is concerned with the existence and availability of methods and knowledge to choose the next task in the design process, execute that task, and select among alternative ways of executing that task to achieve the best outcome. The available knowledge and methods are divided into two categories. *Generative* knowledge and methods are applied to generate new points in the design space, whereas *control* knowledge and methods are used to support efficient convergence of the design process to an acceptable solution.

Then, somewhat analogously to Brown's classification, a task is termed *routine* if sufficient knowledge is available to directly generate the next point in the design space and to converge on a solution with little or no search. A design task is termed *creative* if a problem-solving process is needed to construct the design space in the first place or if the best method available is unguided search through a very large design space.♦ (See sidebar on p. 40.)

Amount of unspecified physical structure. One way of looking at design is as a mapping of intended function into a realizable artifact, device, or structure. Thus, one characterization of a design task is the degree to which the device or structure is left unspecified by the current design task. Different kinds of design tasks that produce physical structure include *structure synthesis tasks*, in which the final structure will be composed from a set of given primitive parts, which may not be resident in the applicable database or knowledge base; *structure configuration tasks*, in which we compose a structure from a collection of specified parts and specified connectors (cf. Sections 4.4 and 4.5); and *parameter instantiation tasks*, in which we obtain values for a set of given parameters.

Difference in abstraction levels between requirements and specifications. We have already remarked on the fact that what is wanted in or from a design is usually stated much more abstractly than are the final fabrication specifications (cf. Chapter 3). Thus, we may have to work down through several levels of abstraction, refining the details each step of the way, as we solve a design problem. The greater the difference between the abstractions of the client requirements and the fabrication specifications, the more complex is the design problem because the greater generality of the former implies a larger space of possible design solutions.

Complexity of subproblem interactions. In this category, we try to assess whether a complex design problem can be decomposed into a set of relatively independent (and easy) subproblems, or whether we need to look for a combination of solutions to subproblems that are highly interdependent. Obviously, the more complex the interaction, the more complex the design problem. Two kinds of interactions are distinguished in this characterization. *Compositional interactions* arise when some combinations cannot be implemented because of *syntactic* differences; that is, the output of a serial device in a circuit cannot be the input to a parallel port of another element. *Resource interactions* arise when combinations of parts or subsystems lead to different global requirements for the (total) system or device being designed. Typically, compositional requirements are relatively local in their effect because the constraints imposed (e.g., by connection rules or syntax) effectively operate only on the small group of objects whose composition defines the current subproblem. Resource interactions tend to operate globally – that is, as a constraint over the resources of the (total) system or device.

Amount and type of knowledge provided by the user/designer. In this analysis, we could argue that human designers are themselves knowledge sources. Thus, a classification of design tasks (especially in terms of whether they are routine or creative, in the first dimension) would reflect the total knowledge provided by both the designer and the design system being analyzed.

Within this general set of dimensions, Tong and Sriram then define four models of routine design, reflecting in each a consideration of both the kinds of knowledge available (and, implicitly, we will see, the representation of this knowledge) and the way the knowledge is applied. These models are:

Conventional routine design involves the application of conventional methods for problems where knowledge-based techniques are not appropriate. For example, when design tasks can be posed in terms of finding the optimal value of an objective function cast in terms of a linear combination of real-valued variables and subject to a set of linear constraints, then *linear programming* techniques (a subset of operations research (OR)) are the appropriate methods to apply.

Knowledge-based routine design is knowledge-based search that includes several different kinds of operations, such as *refinement*, in which we attach more detail to points in the design space; *constraint processing*, in which we prune out alternatives that violate given constraints; and *patching*, in which we attempt to fix or improve incorrect or suboptimal solutions.

Noniterative, knowledge-based routine design refers to design tasks for which there is sufficient knowledge to complete the tasks in one pass through a top-down, increasingly detailed refinement.

Iterative, knowledge-based routine design is required when a single pass will not suffice, which usually happens when there are multiple operative constraints or objectives, in which case various kinds of iteration are required. We return to such iteration in Chapter 6 (see also Section 3.5), but for now we note that these iterative approaches include *backtracking*, in which we attempt to undo errors by undoing their causes through abstraction; *optimization* or *patching*; and *problem restructuring*, in which the designer (typically) suggests a modification in the basic design problem that allows a solution to be achieved.

With these characterizations of design tasks and models of routine design in hand, we could then categorize some of the extant knowledge-based design systems. However, because our interest is less in categorizing software than in understanding design thought processes, we leave further discussion of this analysis of design tasks to the next (and last) section of this chapter.

4.8 Toward a Unified Taxonomy?

Having reviewed these several taxonomies, we face the obvious questions: Can we integrate the ideas contained in these taxonomies into a unified taxonomy of design? Moreover, is it worth it? There are no simple or obvious answers to these questions. We could argue that the endeavor is worthwhile to the extent that a unified taxonomy helps us to understand and articulate that part of design that can be modeled as a set of thought or cognitive processes. This could bring us closer to a coherent theory of design; it could also help us learn (and then teach) what it is that designers do. Furthermore, a unified taxonomy could help us classify computer experiments in design modeling, which in turn helps us externalize our thinking about design and aids in the development of computer-aided design tools that have significant practical application.

However, it is also true that we are put off by the notion of reducing the ideas discussed earlier to entries in a table. A table may be the natural representation for summarizing design-process characteristics, but we are not unsympathetic to the quote attributed by Steier (1993) to the late Allen Newell to the effect that "you can't play twenty questions with Nature and win." In other words, it may be difficult to characterize Nature in terms of a specific set of predetermined categories. Thus, we present what seem to be the most salient ideas about characterizing design in an orderly but nontabular form. We try to formulate taxonomic ideas for both design problems and design processes, but we ignore the design environment. It may be that the design environment is an important issue for locating design work, designing design organizations, or designing design software. However, we assert that as a global principle, the institutional environment within which design is done has little to do with its basic thought processes. However, once a model of design thought processes is in hand, the activities that situate design in any particular environment can proceed more intelligently.

We wish to make one more observation before going further. One of the troublesome aspects of dealing with the design literature is that words are used by different authors to mean different things. For example, design *tasks* are often described as if they were processes, as opposed to the dictionary definition, in which a *task* is "an assigned piece of work" or "something hard or unpleasant that has to be done." Furthermore, in the design literature, methods may include tasks and procedures, whereas in other settings, tasks incorporate methods and techniques and operations. In the sequel, we will set out a characterization of the design process with an explicit ordering of – in order of decreasing abstraction (or increasing refinement) – task, strategy, method, and mechanism. We will try to conform to some degree to the dictionary definitions of these terms that are so critical to successful characterization.

With this in mind, we propose to start by characterizing a design *task* along the lines of the definition of a design *problem* as proposed by Dixon et al. and extended by Ullman. We consider that design can be viewed as a *transformation* of an initial knowledge state to a final knowledge state, much as we described in Sections 4.2 and 4.3. Thus, we need to articulate both the initial and final states, and the transformation(s) by which we proceed from one state to the other. Because the initial and final states are actually representations of the artifact being designed expressed at different levels of abstraction, our state descriptions must be complete and unambiguous at their respective levels of abstraction. Thus, at the start of the design process, we need a sufficiently clear description of the intended end point of the process and the constraints within which the designed device must operate. For the resulting design to be accepted as complete, we must have a set of fabrication specifications that allow the artifact to be built exactly as intended by the designer.

In general, we view the design activity as one of refinement, in which the initial state is more abstract than the final state. Although there might be local variations within a complex design process (e.g., to achieve a specific subgoal, it might be useful to backtrack to a higher level of abstraction to search for other possibilities in terms of physical principles, embodiments, components, and so on), the general direction of design transformation is toward increasing detail or refinement. Furthermore, recognizing that the six kinds of states are themselves ordered (to some extent and not altogether accidentally) according to the degree of refinement, we can perhaps assert that the degree of difficulty of a design problem is roughly proportional to the number of different layers between the initial and final states. Thus, we posit that the initial and final states of knowledge are characterized as each being within one of the following six knowledge-state layers (cf. Section 4.3):

> *Layer 1 – perceived need*
> *Layer 2 – function*
> *Layer 3 – physical phenomenon*
> *Layer 4 – embodiment*
> *Layer 5 – artifact type*
> *Layer 6 – artifact instance*

The next step is the detailing of the knowledge-state layers. Again, we largely retain the structure proposed by Dixon et al. and Ullman. One modification we suggest is a refinement of the values that can be inserted into the representation slots when numerical representations are identified. It may be useful to distinguish between discrete and continuous representations because choices of methods and mechanisms clearly hinge on the nature of the numerical representation. For continuous variables, some parametric or algorithmic approach might be appropriate, whereas discrete variables might indicate that approaches based on selection are more relevant. Thus, five *representation languages* would be identified here (cf. Sections 4.3 and 5.1):

textual
numerical–discrete
numerical–continuous
graphical
physical

The knowledge-state layers and their representations are not likely to be entirely independent because the more abstract the layer, the more likely that it is rather vague knowledge expressed in text.♦ At the other end of this spectrum, artifact types and artifact instances are increasingly specific descriptors. Note that this has a flavor similar to one of the dimensions used by Tong and Sriram to classify design tasks – that is, the identification of the differences in the abstraction levels of the design-problem specification and the implementation of the final device. Thus, this dimension seems to map rather well into the identification of a design problem.

If – and it is a big "if" – the design problem is then set out as a statement of the differences between the (given) initial state and a (sought) final state, we might think of the design process as means–ends problem solving (cf. Section 3.5) and the design *task* is the elimination of the gap between the two problem states by transforming the initial state into the final state. In this broad context, the *ends* are thus the elimination of the differences between the initial and final states, whereas the *means* of achieving that end is an aggregation of a set of subtasks that

There are several reasons for a representation to be "vague," the main one being that design details are unknown at that point in the process. Another possibility is that some of the missing knowledge is "tacit," or not easily expressed, where the owner may not even be aware of knowing it (Haug 2012). Or, it could be that the maker of the representation is making assumptions (Brown 2006) about what knowledge is common and does not need to be expressed. Some or all of the design may have been communicated by sketching (Kara and Yang 2012), where vagueness is often a benefit. Note that such pen or pencil strokes might also act as "gestures" (Visser and Maher 2011) that select or emphasize parts or properties of the design for other design team members that do not normally get captured in formal design representations. More detail about the design intentions and rationale behind early, vague designs can be obtained by using "protocol analysis" (Ericsson and Simon 1993). Protocols are detailed records of the activities of a designer or design team, often including communication in the form of speech, sketches, or gestures. Many protocols are developed by asking the participant(s) to "think aloud": this reveals more of the design process than would normally be available. Collected protocols, such as from the Delft Protocols Workshop (Cross et al. 1996), are analyzed using encoding schemes to learn about the design process and knowledge used (Cross 2001). One scheme annotates each action with a label that indicates whether it is concerned with structure, behavior, or function, or a transition among them (Gero et al. 2011).

make use of domain knowledge including both generative and control knowledge, strategies, problem-solving methods, and various specific mechanisms. Along the way, we will have to deal with a diversity of representations, so we will have to invoke – at least implicitly – translations for moving between different representations. We will elaborate these tasks, subtasks, strategies, methods, and mechanisms herein; for now, it is useful to think of them as similar to but somewhat more abstract than the ways they have been defined previously. At the descriptive levels outlined in Section 3.2, specific design problems can be stated in terms of differences between paired initial and final states, as we have already done in a few cases, and design processes can be elaborated by articulating the design task that needs to be done in terms of the means or subtasks applied in a specific design context.

We now elaborate the characteristic dimensions of design processes, making use of some of the tasks, subtasks, strategies, methods, and mechanisms that we now define (and summarize in Figure 4.9 at the end of the section):

Jackson (1999) includes the often-cited but flawed list of expert reasoning tasks from the 1980s: interpretation, prediction, diagnosis, designing, planning, monitoring, debugging, repairing, instruction, and controlling. Because designing might include some of the other items in the list (e.g., planning), researchers such as Clancey (1985) worked to clarify these tasks into an analytic set (i.e., monitor, diagnose, predict, control) and a synthetic set (i.e., specify, design, assemble), with "configure" and "plan" as subtasks of design. Once such categories are established, it is easier to focus on identifying suitable methods. Although these categories may reflect the general nature of important subtasks, allowing us to identify "a configuration problem," it is better to think of design problems generally as a mix of tasks requiring a mix of methods.

Design tasks.[♦] Thinking of a task as "something to be done," it may be useful to think of design tasks as being the elimination of the difference between initial and final states that can be done relatively easily, with manageable difficulty, or with a great deal of difficulty. In less colloquial terms, design tasks can be characterized as *routine*, *innovative* (or variant), and *creative*. The degree of difficulty (or "routineness"), as noted several times already, depends on the completeness of the knowledge we can bring to bear. That knowledge can be characterized in turn as being either *control knowledge* (i.e., knowledge about managing the process so as to ensure its convergence) and *generative knowledge* (i.e., knowledge used to generate new points in the design space). In addition to our assessment of the available knowledge, there are several other factors that contribute to our view of whether or not a design task is difficult.

One obvious indicator (as noted in the foregoing description of the initial and final knowledge states) is the *difference in the abstraction levels* used to describe these two states. Another indicator of the degree of difficulty is the *degree of interaction among subproblems*, although one could suppose that this is subsumed within the category of the available knowledge. Even where there are many complicated interactions, they present difficulties "only" to the extent that the knowledge available for handling them is unknown or uncertain. Thus, it would seem reasonable to view this dimension as similar to that of identifying the available control and generative knowledge.

It is also interesting to examine design tasks in the context of a well-known spectrum of problem-solving tasks, which ranges from *classification* or *derivation* tasks at one end of the spectrum to *formation* tasks at the other. In classification tasks, a solution is derived from a given set of data or facts. In formation tasks, a problem is solved by forming or synthesizing either an object or a set of plans to create an object. However, whereas design is often assumed to be a formation task, the truth is more complicated; that is, it may be an oversimplified assertion to equate design with formation.

First of all, humans clearly have trouble generating – on demand – creative new solutions for complex, open-ended problems. Design would be difficult to do, therefore, if it involved only the continuous formation of unique solutions. In fact, design tasks at every stage – conceptual, preliminary, and detailed – often involve a style of problem solving that combines elements of both selection (or classification) and formation. Furthermore, we have already seen that several important design tasks involve what have been previously called, respectively, configuration selection (cf. Section 4.4) and configuration arrangement (cf. Section 4.5). Thus, it is probably more meaningful to view design as encompassing both classification (or selection) and formation but also reflecting a desire to achieve a *synthesis* that can then be *analyzed* and *evaluated* (cf. Section 3.3).

This last point leads quite naturally to the last indicator of the degree of difficulty of a design task – that is, the extent to which the physical structure of the designed artifact must be *configured* from a given collection of parts and connectors or *composed* (or *synthesized*) from a collection of primitives. With due regard to the ambiguity of some of these terms and their application, we do recognize that the more we depart from selection and configuration, the greater the difficulty we face in completing a design.

Design subtasks occur naturally in the decomposition of complex design tasks wherein, for example, some parts of a complex system may be designed in a routine fashion, whereas other parts (or the composition of familiar parts) may require more innovation. Thus, design subtasks reflect the recursive or iterative nature of design.

Domain knowledge defines an area of expertise that is used to formulate a response to the posed task. It could include aspects of the knowledge used to generate points in the search space as well as control the application of that knowledge.

Strategies are "careful plans or methods." Strategies are exercised at the highest level needed to properly assess the gap between the initial and final knowledge states. This assessment would account for the differences in the knowledge-state layers, the availability of domain knowledge, the availability of methods and mechanisms (see the following discussion), and, perhaps, the resources available to solve the design problem posed. The availability

Problem-solving methods (PSMs) (Brown 2009; Fensel 2000; Motta 1999) describe how an expert or a computer program can reason using knowledge, and often heuristics, to achieve a goal. Examples of reasoning tasks for which PSMs exist include heuristic classification (useful for selection problems), configuration, and parametric design. A task is usually specified by its goal, its inputs, and its outputs. Several methods might be appropriate for each task. A PSM may solve the problem directly or may decompose the task so that other PSMs can be used. A control structure determines the inferences made by a PSM. The knowledge needed and the *role* that the knowledge plays is also specified. Parametric design specifies the values of a given set of parameters in response to given requirements. Appropriate PSMs include "propose and backtrack" and two types of "propose and revise": "complete and revise" and "extend and revise." The difference between the two is that the former produces a complete design before making heuristic revisions, whereas the latter makes revisions at any point.

(and perhaps cost) of resources, although in part an "environmental concern," could clearly influence the choice of strategy (of which choices have been examined, for example, within the context of structural analysis). *Design planning* is an implementation of strategic thinking (cf. Section 4.3).

Problem-solving methods are procedures or processes or techniques to achieve an objective that "operate as higher-level procedures that organize the application of the mechanisms." Some kind of action must be taken to execute the strategy just identified, and problem-solving methods are those kinds of action. These methods might thus be said to be implementations of the strategies used to solve a given design problem. Ullman's *processing actions* are examples of such problem-solving methods in that they define specific techniques for generating and evaluating points in the design solution space.♦

Mechanisms are "procedures that operate procedurally on well-defined inputs to produce equally well-defined outputs." Examples of these in the domains of configuration design and process design have already been given (cf. Section 4.6). Ullman's *failure actions* (cf. Section 4.3) also may be seen as mechanisms. The proper sequencing of mechanisms is done by the application of control knowledge.

Effects. Although sometimes seen as things that are done, effects are more properly seen as results, perhaps of having applied a method or a mechanism. Some of the identifiable effects of applying methods and mechanisms are (cf. Section 4.3) *refinement*, *decomposition*, and *patching*.

This completes our list of dimensions by which design tasks can be analyzed and described. Although not a formal unified taxonomy, it is an attempt to build on the work already described in this chapter. We will get some indication of how useful it is when we discuss the representation of the design process in Chapter 6. In the meantime, it might best be seen as an informal glossary of terms the formalization of which could be helpful in describing how we proceed thoughtfully through a design process (Figure 4.9).

Design tasks are something to be done; they can be *routine*, *innovative* (or variant), or *creative*; degree of difficulty (or "routineness") dependent on completeness of knowledge.

Design subtasks occur in decomposition of complex designs; they reflect recursive, iterative nature of design.

Domain knowledge is the area of expertise of response to design task.

Strategies are "careful plans or methods," exercised at high levels.

Problem-solving methods are procedures, processes, or techniques to achieve objectives; they operate as higher-level procedures that organize application of mechanisms.

Mechanisms are "procedures that operate on well-defined inputs to produce equally well-defined outputs."

Effects are sometimes things to be done; more often seen as results.

Figure 4.9. An outline of the characteristic dimensions of the design process.

4.9 Bibliographic Notes

The dictionary definition of *taxonomy* is from Woolf (1977).

Section 4.1: Arguments supporting the development of design taxonomies appear in Dixon (1987) and Finger and Dixon (1989a, 1989b) to aid the development of a scientific theory of design, and in Ullman (1992a) to aid the development of computer-based design tools. Design taxonomies are found in Balkany, Birmingham, and Tommelein (1991, 1993); Brown (1992); Brown and Chandrasekaran (1983, 1989); Dixon et al. (1988); Tong and Sriram (1992a); and Ullman (1992a). Search for plausible design alternatives is discussed in Mittal, Dym, and Morjaria (1986). The taxonomy described in this section is explicated in Brown (1992) and Brown and Chandrasekaran (1983, 1989). Innovative design is discussed by Brown (1992), Cagan and Agogino (1987), and Tong and Sriram (1992b). The individual and community measures of routineness are outlined in Brown (1991). Variant design is discussed in Dixon and Dym (1986) and Dym and Levitt (1991a).

Section 4.2: The taxonomy of design outlined in this section is due to Dixon et al. (1988) and is motivated in Dixon (1987).

Section 4.3: The taxonomy of design outlined in this section is due to Ullman (1992a). It is based on the TEA design-process model of Ullman, Dietterich, and Stauffer (1988). The "languages of design" are discussed in Dym (1991) and Ullman (1992b). Mathematical optimization for design is presented in Papalambros and Wilde (1988) and *satisficing* is defined in Simon (1981). Definitions of weak and strong search are found in Dym and Levitt (1991a) and Ullman (1992a).

Section 4.4: The taxonomy of design outlined in this section is due to Morjaria (1989). V-belt drive design is modeled in Dixon and Simmons (1984). The configuration of VAX computers is described in McDermott (1981) and Barker and O'Connor (1989). The design of paper-handling subsystems for copiers is described in Mittal, Dym, and Morjaria (1986).

Section 4.5: The taxonomy of design outlined in this section is due to Hayes-Roth et al. (1987); our discussion is adapted from an extended description in Dym and

Levitt (1991a). The argument that prototypes are the very stuff of design is due to Gero (1987).

Section 4.6: Newell (1981) proposed the ideas of knowledge-level analysis, and Steels (1990) uses "components of expertise" to describe domain knowledge. Clancey (1985) identified the generic method of heuristic classification. Chandra-sekaran (1983, 1986), noting that KBESs may not be as flexible as human design-ers, proposed a set of generic design tasks. The taxonomy of design outlined in this section is due to Balkany, Birmingham, and Tommelein (1991, 1992). The results of their analysis were used to guide the construction of a domain-independent design environment (Runkel et al. 1992). The system CPD2-Soar is described in Modi et al. (1992); it is based on the Soar architecture of Laird, Newell, and Rosenbloom (1987). Knowledge types are discussed in general terms in Dym and Levitt (1991b); for applications to structural engineering in Jain et al. (1990) and Luth (1990); and for design in Dym, Garrett, and Rehak (1992).

Section 4.7: The taxonomy of design tasks outlined in this section is due to Tong and Sriram (1992a). The degree to which a design task leaves something unspecified is explored by Steinberg and Ling (1990). A categorization of some KBESs in terms of their taxonomy is also given by Tong and Sriram (1992a).

Section 4.8: Steier (1993) brought Newell's quote to our attention. The dictionary definitions used in this section are from Mish (1983). The definitions we present here represent in some sense an adaptation of the ideas of Balkany, Birmingham, and Tommelein (1991, 1993); Brown (1989); Dixon et al. (1988); Tong and Sriram (1992a); and Ullman (1992a). The well-known spectrum of problem-solving tasks is Amarel's (1978). The role of strategic knowledge in structural analysis is explored by Salata and Dym (1991).

5 Representing Designed Artifacts

We now turn to the representation of designed artifacts. By focusing first on objects, we risk introducing an artificial distinction into the scope and meaning of *design knowledge*. Clearly, design knowledge must incorporate information about design procedures, shortcuts, and so on, as well as about artifacts. Furthermore, there is some evidence that designers think about processes as they begin to represent the objects they are designing, especially when they begin to create sketches and drawings. And, of course, to fully represent objects and their attributes also means being able to fully represent concepts (e.g., design intentions, plans, behavior, and so on) that are perhaps not as easy to describe or represent as physical objects. However, because the end point of most engineering designs is a set of fabrication specifications for an object (and, occasionally, the object itself) and because engineers often think in terms of devices, we start our discussion of representation with physical objects as our focus. As a consequence, we may somewhat overlap our later discussion (cf. Chapter 6) of the representation of design knowledge and design processes.

5.1 The Languages of Engineering Design

To discuss different representations of designed objects or artifacts is to talk about the languages or symbols in which those representations are cast. That is, for us to describe an object, whether real or conceptual, in detail or abstractly, we must effectively choose a language with which to write our description.♦ We pointed out

> Visser (2006) points out that designing means constructing representations: both mental and physical. One of the points of this chapter is that different representations have different strengths. Consequently, it is both necessary and important that we be able to:
>
> - Integrate representations because the multiple aspects of a designed object are interrelated (e.g., behavior depends on structure, surface finish depends on material and on manufacturing processes).
> - Translate one representation into another to enable additional representation-dependent reasoning, faster processing (as in knowledge compilation), and knowledge acquisition (e.g., component knowledge from text), or to perform tool-to-tool information transfer (Eastman et al. 2010).

previously (cf. Sections 4.3 and 4.8) that there are several "languages" in which design information about objects is cast as follows:

Verbal or **textual statements** are used to articulate design projects; describe objects; describe constraints or limitations, especially in design codes (see following discussion); communicate among different members of design and manufacturing teams; and document completed designs.

Graphical representations are used to provide pictorial descriptions of artifacts. These visual descriptions include sketches, renderings, and engineering drawings. The role of CADD systems in producing design pictures of all kinds is evolving rapidly, as we discuss in cf. Section 5.2.

Mathematical or **analytical models** are used to express some aspect of an artifact's function or behavior, and this behavior is in turn often derived from some *physical* principle(s). Thus, this language could be viewed as a **physical representation language** (cf. Sections 4.3 and 4.8).

Numbers are used to represent design information in several ways. Discrete values appear in design codes and constraints (see the following and as attribute values (e.g., part dimensions). Numbers also appear through continuously varied parameters in design calculations or within algorithms wherein they may represent a mathematical model.

We include additional notes in this chapter about a variety of representations, some new and some already mentioned, including functional representations, causal networks, configuration spaces, shape grammars, solid models, voxels, description logics, ontologies, and neural nets.

♦We add to this list two more representations that use words or textual statements but in a very confined format. They derive from programming constructs used within the context of representation for symbolic computing. These representations are, in fact, the core ideas behind AI-based programming (cf. Chapter 1). Their syntaxes or grammars are highly stylized and specific, so they are worth identifying as separate "languages." Moreover, we use them extensively in this and subsequent chapters. These additional representation languages are as follows:

Rules are statements to the effect that we should perform a specified *action* in a given *situation*. Typically written in sets of IF–THEN clauses, the left-hand (IF) sides of the rules define the situation(s) that must obtain before a rule can be applied, whereas the right-hand (THEN) sides of the rules define the actions to be taken. In design, rules are used to represent (1) design heuristics or rules of thumb; and (2) design codes, such as the building codes that many governmental entities enforce on buildings and other structures; we subsequently present other examples.♦

In fact, rules are a universal language that we can use to represent any design process or design requirements. Because this chapter is about representing artifacts, it is clear that rules that state facts that must (or not) be true of an object can be viewed as declarative knowledge. Even some process-oriented rules, such as those that describe possible design-task decompositions, may be interpreted as statements about artifact structure.

Objects or **frames** are structures that can be used to represent *objects* and their *attributes*. They are, in fact, the very heart of what is called *object-oriented representation*. Frames can be viewed as elaborate data structures that can be linked to other frames through a variety of logic-based and/or procedural calls to data or attributes or calculation methods. These objects or frames are interesting because, together with their underlying networks of links and procedures for storing and passing data, they

More recently, description logics (DLs) provide a more formalized frame-like representation, based on logic, with a more prescribed, well-formed, and consistent usage. For example, DLs provide a way to "classify" a new concept so that it can be put into the right place in an existing hierarchy of concepts. Brachman and Levesque's (2004) book provides an introduction to DLs, whereas Baader et al. (2003) is a useful reference. Note, too, that languages such as OWL for the semantic Web (the semantically annotated World Wide Web) are based on DL.

can be used to relate descriptions of physical and conceptual objects to each other in rich and interesting ways.♦

We recognize that classifying these two representation schemes – rooted as they are in symbolic computation – as design languages may be somewhat controversial. However, we will see that these particular constructs do offer effective ways of stating and applying design knowledge. Furthermore, we also will demonstrate that these structures can be used to organize and integrate design knowledge that is represented in one or more of the other design languages, thus laying the groundwork for developing integrated computational environments for design. However, to illustrate their utility for expressing information about designed artifacts, we present here brief examples of both a rule and a frame that are part of a KBES (called DEEP) built to assist designers in configuring electrical service for residential plats. The rule describes how building lots on a street should be clustered or grouped so that they can be configured efficiently for service. It does this by combining within the rule heuristic knowledge (i.e., a judgment about how close the lots on a street are to one another) with a relationship derived from basic electrical-circuit principles (i.e., the transformer size required to meet a specified service demand).

IF the number of lots on the street is very close or equal to the maximum number of services that a 50 kVa transformer can serve

THEN cluster all of the lots on that street and serve them with a 50 kVa transformer.

The frame example we show here (Figure 5.1) illustrates how particular examples of a kind of object, called *instances*, can be described in relationship to a *class* of objects that share common attributes. We show in our example an instance of the class of transformers. (As with the rule example, there is much more complexity than is described here because the actual links between the instance and its classes are not shown.) The object description contains slots for particular attributes (i.e., the left-hand column in Figure 5.1) and it contains (in the right-hand column) the specific

SLOT/ATTRIBUTE	VALUE
SuperClass	Components
Class	Transformers
InstanceName	XFR50_1
Type	BURD
DefaultType	PADMOUNT
Rating (kVa)	50
Size (ft^2)	48
Structure	CONCRETE PAD
MaxConnections	6
Coordinates	(X1, Y1)
MaxCustomers	21

Figure 5.1. The instance XFR50_1 of the class Transformers (Demel et al., 1992).

values for those attributes that distinguish the particular transformer in question (i.e., XFR50_1. The slot for the attribute DefaultType is *inherited* or passed down from the class Transformers, here as a PADMOUNT transformer, unless it is overridden at the local level either by a rule application or through intervention by the system user. In addition, the links (not shown here) can be used to establish and maintain various kinds of relationships; for example, a CONCRETE_PAD may be *part-of* a PADMOUNT transformer, whereas a BURD is a *kind-of* transformer.

Now, there are other constructs that we could extract from symbolic programming, in particular the styles called *logic programming* and *neural networks*. However, we choose not to identify these as design languages or representations for the following reasons. Logic programming is an implementation of predicate and propositional calculus that has, in our view, severe limitations in terms of its expressive power. We also would argue that rules, as an extension of logic programming, serve our purposes quite well. Similarly with regard to neural networks (NNs),[♦] a much newer computational device, neural networks are beginning to show some promise for adaptive systems that can "learn" as they develop. However, neural networks are much less a representation scheme than they are transformational systems that aid in the learning process.[♦] Perhaps at some point there will be an integration of neural networks with the greater power offered by object-oriented programming, but at least in today's computing environment, they do not have as much to offer.

> Russell and Norvig (2010) include a short introduction to neural networks, whereas Gurney (1997) is a more detailed but introductory alternative.

> Because neural networks (NNs) can classify, it might be argued that, coded in their connections and weights, they contain knowledge about the properties of each class. But although there have been some attempts to extract rules from NNs, their main application has been as situation recognizers.

As this work unfolds, at different times we employ different languages to represent design knowledge. Furthermore, we cast the same knowledge in different languages to serve different purposes, which in turn require that we link the different languages as we represent different aspects of an artifact or different phases of the design process. Consider for a moment the structural-design problem outlined in Chapter 1. We noted that the complete design of a structure requires many kinds of knowledge, and that we use several different knowledge-representation schemes. In light of the current discussion, we could say that we are casting various kinds of structural-design knowledge in terms of the different languages we have delineated. For example, fundamental structural-mechanics knowledge can be expressed as follows:

Analytically; for example, formulas for the vibration frequencies of structural columns.

Numerically; for example, in discrete minimum values of structural dimensions or in FEM algorithms for calculating stresses and displacements.

In terms of *heuristics* or *rules of thumb*; for example, that the first-order earthquake response of a tall, slender building can be modeled as a cantilever beam whose foundation is excited.

Similarly, knowledge about the equilibrium behavior of certain thin structures can be expressed as follows:

As a *rule*; for example,

IF   a structural element has one dimension much thinner than the other two
AND   it is loaded in that direction
THEN   it will behave as a plate in bending.

Mathematically; for example, the fourth-order partial differential equation that governs the deflections of bent plates.

Verbally and *numerically*; for example, the deflection of a floor in a residential building should not exceed its length (in feet) divided by 360.

We have thus restated equilibrium for a bent plate in several forms. We often cast the same knowledge in different languages, depending on the task at hand. As we have said before, experts manage to choose the right language at the right time to solve the immediate problem. However, we need to continually recognize that we are employing different languages, that these different representations offer us different insights and utility, and that it would be desirable (in computational terms) to link these different languages so that we can seamlessly model our designed artifact and our design process.

5.2 Images of Designed Artifacts

We now turn to the role of graphical or pictorial languages, by which we mean to include sketches, freehand drawings, and CADD models that extend from simple

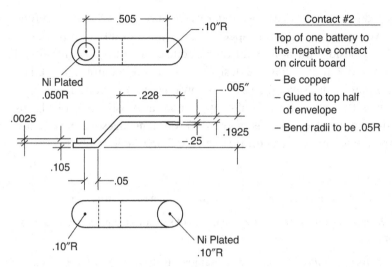

Figure 5.2. Design information adjacent to a sketch of the designed object (Ullman, Wood, and Craig, 1990).

wire-frame drawings through elaborate solid models. Our discussion is strongly influenced by a recent study of the importance of drawing in the process of mechanical design because we think the lessons learned from that study are widely applicable to engineering design. We, too, will focus largely on the roles of the images, leaving for separate discussion issues relating to the media or environments in which images are created. For example, we discuss in Section 5.6 the value of combining visual and other data through their various representations, so some media-related issues are discussed there. For now, we wish to concentrate on the information transmitted in the drawing process.

Historically, we are talking about the process of putting "marks on paper." Researchers have observed in their investigations that these marks include both graphic representations and support notation. The graphic representations include sketches of objects and their associated functions, as well as related plots and graphs. The support notation includes notes in text form, lists, dimensions, and calculations. Thus, the marks (or drawings) serve to facilitate a parallel display of information because they can be surrounded with adjacent notes, smaller pictures, formulas, and other pointers to ideas related to the object being drawn and designed. That is, a sketch or drawing offers through adjacency the possibility of an organization of information more powerful than the linear, sequential arrangement imposed by the structure of sentences and paragraphs. An example that illustrates some of these features is shown in Figure 5.2, which is a sketch made by a designer working on the packaging – to consist of a plastic envelope and the electrical contacts – to accept the batteries that provide the power for a computer clock. Here, the designer has written down some manufacturing notes adjacent to the drawing of the spring contact. Furthermore, it would not be unusual for the designer to have scribbled modeling notes (e.g., "model the spring as a cantilever of stiffness . . . ") or calculations (e.g., calculating the spring stiffness from the cantilever beam model) or other information relating to the unfolding design.♦

Now, marginalia of all sorts are familiar sights to anyone used to working in an engineering environment. In doing both analysis and design tasks, we often draw pictures and surround them with text and equations. Conversely, we often draw sketches in the margins of documents, perhaps to elaborate a verbal description, perhaps to indicate more emphatically a coordinate system or sign convention. Thus, it should come as no surprise that sketches and drawings are as essential to engineering design as any other representation. Interestingly enough, whereas some of the classic engineering design textbooks stress the importance of graphical communication, the topic seems to have vanished in more recent works.♦ In domains such as architecture, of course, sketching, geometry, perspective, and visualization are acknowledged as the very underpinnings of the field.

There are many major issues relating to graphical representation, some of which are less important in the current context. For example, graphics can be seen as a very important aid to memory and as a means of communicating between short- and long-term memory, as we discussed when we described an information-processing model of design in Section 3.4.♦ In a similar vein, graphic images can be used to communicate with the external environment – that is, with other people – so such images are important in design and manufacturing organizations (see Section 5.7). Perhaps it is useful to recapitulate here a partial list of how drawings are used in the design process:

Different models of the same object or mechanism may exist, depending on the assumptions that they make (e.g., no friction). Models can be related by the assumptions that they add or remove, producing a "graph" of models. When models exist only for components rather than for the whole product, product analysis or simulation cannot be done until a complete product model is assembled from compatible, smaller existing models of its parts (Nayak et al. 1991). This may well be task dependent: modeling a current-heated wire as a heat source is appropriate for analyzing operating temperatures, whereas modeling it as a spring would not be.

Although sketching is still included in some university coursework – at Massachusetts Institute of Technology and Imperial College, for example – there is little mention of it in recent engineering design texts, although Dym et al. (2009) describe using different types of sketches to convey design information. Sketch recognition for mechanical and electrical design is an active area of research. The goal is to correctly and automatically recognize design intent from sketches and convert it into a representation that can be used computationally. For example, see the work by Gross, Hammond, Alvarado, Kara (Fua and Kara 2011) and by Stahovich (Bischel et al. 2009).

Graphical representations can also act as a form of design rationale, where sketches can show the development of an idea or be the cause of a design decision (e.g., interference between parts might be seen). Even nongraphical pen or pencil movements can be seen as gestures that convey meaning and might act as rationale.

1. To provide a permanent record of the shape or geometry of a design.
2. To facilitate the communication of ideas among designers and between designers and manufacturing specialists.

3. To support the analysis of an evolving design.
4. To simulate the behavior or performance of a design.
5. To help ensure that a design is complete because a picture and its associated marginalia could serve as reminders of still-undone parts of that design.

Ullman, Wood, and Craig advance several hypotheses about the role of drawing in the evolution of a design and as a part of the cognitive process of design. Some of these hypotheses are particularly relevant to our concerns and the first such is as follows (our ordering differs from the original ordering of the authors):

> "Drawing is the preferred method of external data representation by mechanical engineering designers."

One type of representation for material breaks the volume of an artifact into many small cells, usually called "voxels" or "elements." Although this sort of representation is commonly used in engineering to calculate stresses in an object (e.g., finite element analysis), evolutionary algorithms from AI offer the possibility of modifying the shape and/or the material to produce artifacts that are lighter, stronger, better performing, or easier to manufacture. Sims' (1994) work influenced a lot of research by demonstrating evolutionary algorithms that developed designs for small geometric "creatures" whose fitness (e.g., how well they could move) could be tested in a simulated world. Pollack et al. (2003) report work on structures and on robots, but with fitness tested by actually making them. The voxel representation also has been used for topology design, taking stress and material distribution into account (i.e., less material where it is not needed) (Jakiela et al. 2000). Kicinger et al. (2005) provide a survey of evolutionary computation used for structural design.

Perhaps this is no more than a reflection of what Woodson cites as the more accurate translation of a favorite Chinese proverb: "One showing is worth a hundred sayings." But perhaps it is also a reflection of a German proverb also quoted by Woodson; that is, "The eyes believe themselves; the ears believe other people." In fact, a good sketch or rendering can be very persuasive, especially when a design concept is new or controversial. And, as noted previously, drawings serve as excellent means of grouping information because their nature allows us (at least on a pad or at a blackboard, if not yet in a CADD program) to put additional information about the object in an area adjacent to its "home" in a graphical representation of the object being designed. This can be done for the design of a complex object as a whole, or on a more localized, component-by-component basis. Furthermore, graphical representations are effective for making explicit geometrical and topological information about an object.[♦] However, pictorial representations are limited in their ability to express the ordering of information, either in a chain of logic or in time.

Still another hypothesis of interest is that:

> "Drawing is a necessary extension of visual imagery used in mechanical design. It is a necessary extension of a designer's cognitive capability for all but the most trivial data representation, constraint propagation, and mental simulation."

The argument made here is that some external, graphical representation, in whatever medium, is an absolute prerequisite for the successful completion of all but the most trivial designs.♦ Both the research underlying this hypothesis and our everyday experience – think of how often we pick up a pencil or a piece of chalk to sketch something as we explain it, whether to other designers, students, teachers, and so on – support this notion and probably more so in mechanical engineering design than in some other domains. This is perhaps because mechanical artifacts quite often have forms and topologies that make their functions rather evident.♦ Think, for example, of such mechanical devices as gears, levers, and pulleys. This evocation of function through form is not always so clear, and sometimes more abstract graphical representations are used to show functional verisimilitude without the detail of sketches that are based on physical forms. Two examples of this sort of graphical abstraction are the use of (1) flowcharts to represent chemical-engineering-process plant designs, and (2) block diagrams (and their corresponding algebras) to represent control systems.♦ It is likely true for all domains, with the varying kinds of abstraction levels we have just seen, that these pictures and charts and sketches serve to extend our limited abilities, as humans, to flesh out complicated pictures solely within our mind.

The final hypothesis of interest here is:

"Sketching is an important form of graphical representation serving needs not supported by drafting."

The issue here is the relationship of quick and informal sketches to more formal drawings whose preparation may require considerably more time and resources.

Heylighen (2011) discusses the role of sketching, and of sight in general, in her study of a blind architect. The architect uses models in clay, or sometimes Lego, as well as verbal communication. He "sees" both the models and the actual buildings with his hands. Athavankar et al. (2008) studied how blindfolded designers and architects use hand gestures to make their mental designs and images visible to others. He also found that blindfolded architects could physically navigate an empty space, imagining themselves in an architectural space that they have designed. Bilda et al. (2006) carried out a study that contrasted the designs of expert architects who were allowed to sketch with those who were not. Surprisingly, they found no significant difference between the two groups and so concluded that sketching is not an essential activity during conceptual design.

Designers and design researchers commonly say that there is "no function in structure," meaning that in order to have a function, an artifact has to be placed in a context and produce a desired effect. The placement, context, and "desire" all can be varied (e.g., a pen used as a pointer). It is worth noting that "no structure in function" also applies because a function is always stated in an implementation-free manner (e.g., a device that "provides pointing" could be a stick or a laser).

Causal representations have been used in AI for quite some time. Knowing that A causes B allows a system doing diagnosis to suspect a problem with A if B does not occur when it should. Such reasoning can be useful during design because design decisions lead to a device having a behavior; but, if that behavior is not what is needed, then some design decision (or decisions) might be wrong. We also can use causal representations

(continued)

(*continued*)

to represent "behavior" in function-behavior-structure representations, allowing qualitative simulation to be done. More recently, Bayesian networks (Charniak 1991; Russell and Norvig 2010) have become increasingly popular, with causal dependencies among events quantified by conditional probabilities. Giving evidence about any subset of events allows us to compute the probabilities about any other subset.

Schön (1983) refers to designing as a "reflective conversation with the situation," where the situation might include a sketch. This enables the designer to reinterpret it and so be reminded of other possibilities. The four main activities Schön proposes are *naming* the relevant aspects of the situation, *framing* the problem in a particular way, *moving* toward a solution (i.e., making design decisions), and *reflecting* to evaluate the "moves." Visser (2006) provides a detailed set of the critiques of Schön's ideas: essentially that they show "lack of precision" (i.e., they do not provide enough detail to allow the design of a computational model). Gero and Kannengiesser (2008) remedy this by casting Schön's ideas in terms of their situated function-behavior-structure model, providing a detailed description of "reflection."

Configuration spaces are a representation intended to show the "space" of allowable relative positions of a pair of objects. For example, the center of a robot moving among obstacles can only be in certain places in the space: this could be mapped in a diagram with two dimensions (x, y), showing regions allowed and not allowed. For an object pushing against a rotating cam, the object moves toward and away from the axis around which the cam is moving. We then obtain a picture of allowable positions by plotting angle

(*continued*)

Based on reported observations of architectural designers at work, this hypothesis suggests that sketches have a unique role because they are readily created and easily changed.♦ We can see, for example, that sketches make it easy for a designer to play with ideas and concepts in much more of a "brainstorming" mode than is possible with formal drawings or blueprints. And, because they can be made in the margins of blueprints as well as in other contexts, informal sketches serve both as intermediaries between and as "graphic metaphors" for the formally drawn plans that depict artifacts as well as for artifacts themselves. We should recognize here that this observation raises interesting questions about the way both sketching and drawing are taught (or not, as is increasingly the case) in engineering programs, as well as about the future development of CADD systems. We return to some of these questions in Section 7.3.

For now, we close this section by observing first that there is a very strong case to be made for the fact that graphical representations of all kinds – whether done by hand or on a computer, whether informally sketched or detailed in a complex set of blueprints, whether presented bare or annotated with marginalia – represent a great deal of design information. As already noted, the importance of graphical representations in engineering design has been known for some time. What is of particular interest to us is the leverage gained from integrating such representations with other kinds of representations.♦ We have more to say about this in Section 5.6, as well as in the concluding chapter.

Our final observation is that although we have used the word *images* several times in this section as well as in its title, we have not made any reference to photographic images (or, for that matter, have we referred to the "graven images" of the Ten Commandments – and we will not start now!). Certainly, photos have much of the content and impact that are ascribed to other graphical descriptions, but they do not seem to be widely used in engineering design.♦ One possible exception may be the use of optical lithographic techniques to lay out very large-scale integrated (VLSI) circuits, wherein a photography-like process is used. However, it is also true that we are increasingly collecting more data by photographic means (e.g., geographic data obtained from satellites). With computer-based scanning and enhancement techniques, we should expect that design information will be represented and used in this way. One sign of this trend is the increasing interest in geographic information systems (GIS), which are highly specialized database systems designed to manage and display information referenced to global geographic coordinates. It is easy to envision that satellite photos will be used together with GIS and other computer-based design tools in design projects involving large distances and spaces (e.g., hazardous-waste disposal sites and inter-urban transportation systems). Thus, we should not forget photography as a form of representation.

(*continued*)

of rotation against axis-to-object distance. This sort of reasoning allows configuration spaces to be constructed for mechanisms. Configuration spaces support retrieval of stored mechanisms (Murakami 2002) and tolerance analysis, as well as enabling a designer to manipulate the diagram of the space to point out where the mechanism needs to be changed (Joskowicz and Sacks 1997).

Photographs, as well as sketches, plans, schematics, or technical drawings, can act as cases for use in case-based reasoning, providing examples and detailed information for designers. The Web provides access to a huge number of images, so it can serve as a case base. Image matching is now possible, allowing various kinds of image-driven Web searching. Image search by concept map (i.e., spatially arranged text) is possible, where the word *Jeep* placed above the word *grass* could be used to retrieve pictures of Jeep vehicles on grassy ground. Image search driven by a given image is also possible, where features of the given image (e.g., its color distribution) are used in the search. Another kind of image search on the Web is driven by sketching (e.g., the Microsoft MindFinder project). For example, drawing two circles next to each other might retrieve images of cars by matching with wheels.

5.3 Feature-Based Descriptions

We now turn to a representation that can be viewed in some sense as a bridge between graphical representations and the object-oriented descriptions we discuss in the next section. CADD (and originally CAD) systems have been used for some time to draw pictures and plans of objects as they are being designed. The representations in such systems are generally limited to points, lines, and surfaces. Designers, however, think about artifacts in ways that are much more encompassing. Even when thinking about issues of geometry and topology, designers work in terms of aggregated information and concepts – for example, shapes of surfaces and volumes, holes, fit, interference, tolerances, and so on. In other words, even when reasoning about spatial issues, we need a richer and more powerful language or representation to enhance our ability to reason about the objects we are drawing (and designing). *Feature-based representation*, as it is called, has emerged from just such attempts to enable reasoning about objects for which data is stored in CADD systems. Thus, the connection between feature-based and object-based descriptions is that the latter turn out to be a useful vehicle for representing the additional information about features that we cannot represent in terms of the points, lines, and surfaces used in CADD systems.

Features were originally thought of as volumes of solids that were to be removed (typically by a machining process) because of the interest in relating a part to some part of a process plan. More recently, they have been assumed to include devices such as gears, bearings, and shafts, in the contexts of relating an artifact's intended function to a form feature in the CADD representation and of trying to represent the physical features of an artifact so that they can be evaluated. Intermediate to these are hierarchies of features of a geometric nature that, although they enable certain functional reasoning, also can be seen as logical extensions of the original idea of features for processes other than machining. Examples of feature-based descriptions include windows, corners, and tongues for injection molding; walls and fillets for extrusion; and walls and boxes for casting. We will soon see that feature hierarchies are naturally represented in object-oriented descriptions.

What exactly do we mean by a *feature*?[♦] Although used more often and in a larger number of contexts, features are still regarded differently by various researchers and users, so that a definitive definition has yet to emerge. Its first use was in the phrase *form features*, which are particular shapes or volumes associated with a part, such as holes and slots. It was after further work that the use of the term broadened to be inclusive of both form and function, and the context extended to manufacturing and life-cycle concerns of designed shapes and objects in which both geometric and behavioral issues figure prominently. This is largely due to the fact that, as have noted earlier, designers do not think in terms of points, lines, surfaces, edges, and so on.

> Shah and Mäntylä (1995) provide the classic, comprehensive overview of features. Brown (2003) reviews various definitions and provides a version based on the concept that "a feature is anything about the artifact being designed that is of interest to the designer" (p. 895), specifically, anything about the structure, behavior, or function that might affect a goal associated with a particular process. For example, a process might be manufacturing, use, or assembly, affected positively or negatively, whereas a goal might be to be inexpensive, take little time, or be highly useable.

They tend to think in terms of particular forms that are intended to serve a function imposed by the designer. For example, consider the part shown in Figure 5.3. It is used as a locator for a piece that is formed within a mold, and is itself made through the process of injection molding. We must require the shaded surfaces to remain parallel and at a specified distance from one another if we expect the molded piece to be kept at a desired location. Now we focus briefly on the tongue and window, and especially on their functions.

The tongue, as noted, serves to locate the piece within the die. If we want to change some property of the tongue, we want to ensure that any changes are consistent with the locating function. Thus, if we modify the tongue's geometry, we

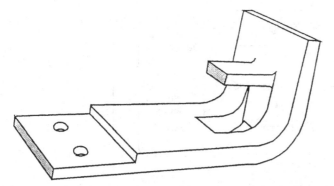

Figure 5.3. An injection-molded part (Dixon, Libardi, and Nielsen, 1989).

want to enforce corresponding changes in the rest of the part so that we do not inadvertently defeat our intended purpose. Similarly, the window over the tongue makes it easier to mold the piece being molded within a two-plate die whose parting direction is normal to the tongue, and it makes it easier to extract the part from the two-plate die. (The parting plane is the plane between the two die faces; the faces are separated or parted in a direction perpendicular to the parting plane.) If, therefore, we want to change the nature of the window, we must make sure that we do not defeat the two purposes that the window serves in our design.

Furthermore, in both cases, as we think about our design (as just outlined), we think in terms of the *tongue*, the *window*, and (if we are being careful) the purposes we want these features to serve. We are not thinking about the coordinates that define the window opening or the precise location and orientation of the face at the end of the tongue. We are, in fact, thinking in terms of the features and their meaning in terms of what we expect them to do in the finished design. Current CADD systems, however, typically cannot accommodate these views of the design unless we extend their respective representations beyond the geometric data into the aggregated features that we use when we think about the unfolding design. Thus, as in our previous discussion (cf. Section 5.2) of the role of sketching in design, we again recognize a need to group information and integrate representations.

We content ourselves for now with a brief look at how we might link CADD data to richer descriptions. At the same time, we show how object-oriented representations lend themselves naturally to enhancing applications of features. There are two basic ways to go about this process of linking CADD data to other descriptions. In the first instance, we *extract features* from traditional CADD databases; that is, we try to identify form features in the CADD representation and use algorithms to extract them so that they can be analyzed for their manufacturability. In the second approach, features are identified and designed in systems that serve as front ends to CADD systems. The resulting designs are then stored as features in the associated CADD databases. The front-end systems are often KBESs that are built specifically for this purpose. This approach, sometimes called *design-with-features*, appears intrinsically more interesting because it allows us, as designers, to work with the concepts we use to think about our work and to then translate the resulting design into a CADD database entry or file.

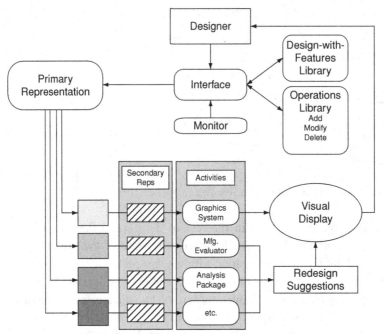

Figure 5.4. An architecture for a computer-aided design system based on design-with-features (Cunningham and Dixon, 1988).

A basic architecture for systems intended to facilitate design-with-features is shown in Figure 5.4. Central to this approach is a library of features called *design-with* features, which are used by the designer to build up a primary representation of the designed artifact expressed in terms of these features. One way that we can characterize a library of design-with features is to sort them into *static features*, the basic function of which is structural, and *dynamic features*, which pertain to movement or the transmission of energy. The second class, which could include devices such as gears and crankshafts, has not been developed as yet even in research systems. The class of static features has been elaborated in terms of five kinds of features: *primitives*, *intersections*, *add-ons*, *macros*, and *whole-forms*. Each class can be elaborated in greater detail by identifying various attributes that take on specific values by assignment, calculation, inheritance from a more abstract class, or some other means (see Sections 5.1 and 5.4). We illustrate two of these classes of static features to make the points that (1) they are at the right level of abstraction for more immediate use by a designer than are basic CADD representations, and (2) they can be sensibly expressed in the object-oriented language of design.

Thus, the primitive features are the primary structural building blocks of a design-with-features design, two of which are solids and walls, the further decomposition of which is shown in Figure 5.5. Add-ons are features added to primitives to achieve some local effect or function (e.g., *depressions* and *openings*). Macros are combinations of primitive features that are structured in advance (e.g., *wall combinations* and *wall-solid combinations*). Intersections are features that provide a syntax for combining either primitives or add-ons on a pairwise basis. Some of the intersection features are illustrated in Figure 5.6. We can see from Figures 5.5 and 5.6 that

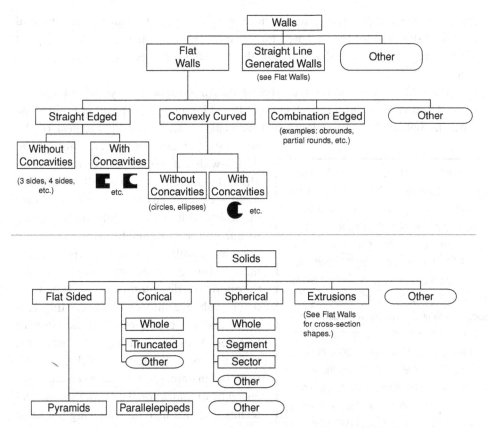

Figure 5.5. Primitive features (Dixon, Cunningham, and Simmons, 1989).

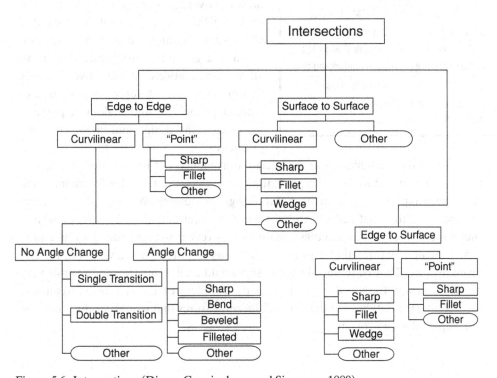

Figure 5.6. Intersections (Dixon, Cunningham, and Simmons, 1989).

the features are expressed in terms, here mostly geometric, that are at a natural level of abstraction for the designer. Furthermore, we also can see the beginnings of the hierarchical structure that is one of the hallmarks of object-oriented representation.

A complete review of feature-based design systems is beyond our current scope. However, our brief introduction to this still-emerging concept not only illustrates its potential; it also illuminates its weaknesses. First and foremost of the latter are the extent and generality of the design-with features library. It is rather unrealistic to assume that we will be able to build – at least any time soon – a library of design-with features sufficiently robust to encompass completely even one design domain, together with all of its relevant manufacturing and assembly processes. Furthermore, as designer/users, we will want to (1) enhance and combine existing features, and (2) create our own design-with features. Thus, we will have to confront a variety of issues, some of which are more technical in nature, others of which are influenced by organizational considerations. The technical issues include balancing the need to have enough features to provide comprehensive coverage against the combinatorial explosion that could emerge if we articulated all possible features and their attributes, creating a grammar sufficiently flexible to allow combinations of primitive features that might be very different than their constituent primitives, and making it possible for us to easily design and create our own primitives.♦

> Shape grammars are similar to grammars for natural languages such as English (Brown 1997). Instead of a rule such as "a verb phrase is a verb with a noun phrase," a shape grammar might have a rule such as "a house is a block with a roof," where shapes – rather than words – are the primitives. We can then describe complex designs by providing a set of rules that describe modifications to 2D or 3D shapes. Theoretically, we can use such rules to analyze the structure of an existing design (analogous to parsing a sentence) or to generate (synthesize) one design, or all possible designs, from a given shape "seed." We can attach actions to rules to detect the correctness of the resulting shape, or rule use can give values to attributes (e.g., volume). One of the strengths of this representation is that unlike a human designer, it does not become fixated on a particular approach or biased by past designs. This allows portions of the design space that normally would not have been considered to be generated and presented to designers, thus encouraging more diversity and increasing the potential for more creativity (Cagan 2001).

One of the challenges – and associated problems – of feature development is that we have in feature-based representation the nucleus of an idea for maintaining different perspectives of an artifact as its design is emerging. We could develop form features, features that reflect different analysis requirements (e.g., stress analysis, thermal analysis, and kinematics), features that reflect manufacturing or assembly requirements, features that represent marketing requirements, and so on, for any given artifact. That is, feature-based design could enable the development of very powerful design tools – if we can solve the problem of translating between and maintaining the database of the many different features that would certainly emerge in complex design projects.

The institutional or organizational issues include maintaining consistency across a design organization, maintaining version control, and having a system interface sufficiently friendly to enable us to combine and create features in the design context and vocabulary in which we normally do our work. (And, we add parenthetically, a design-with-features system would be commercially successful only if we, as designers, can learn to use it without having to become systems programmers. After all, one of the very clear lessons of commercial software development is that the success of a computer tool strongly correlates with the extent to which it is perceived as "user friendly." Although we have said repeatedly that our main interest in this exposition is the design thought process and not the resulting software, we are not entirely impervious to the software-related implications.)◆

> We must not forget that human–computer interaction (HCI) is central to dealing with representations and to designing in general. Lee et al. (2010) estimate that about 1 million US workers use computer-aided engineering or design tools, and they review "usability principles and best practices" for such tools. HCI professionals translate "user friendly" into more explicit specifications: testable requirements and sets of metrics, such as time to learn, error rates, and task completion time.

5.4 Object-Oriented Descriptions

We now come to the representation technique that can be said to be the very heart of this chapter and perhaps, therefore, of this book. Because we have referred to object-oriented representation as both a representation technique and a language of design, along with rules, we intend to view object-oriented descriptions through both prisms in this section. We begin by returning to our ladder problem (cf. Sections 1.1 and 3.1) because we use ladder design to illustrate object-oriented descriptions in more detail. In particular, we start with a class of objects called Ladder (shown in Figure 5.7). The physical object is expressed as a collection of attributes or slots. A value is identified for each slot in one of several ways. Values can be specified by the user, found by a prescribed calculation or procedure, identified through a link to one or more different objects, or prescribed by default. Here, the height is an input parameter specified by the user, as is the type of ladder, although only a

SLOT/ATTRIBUTE	VALUE
Height	User must specify
Sub_classes	(Step Folding Extension)
Width	0.15 * Height
Depth	0.02 * Height
Rung_count	Height [in]/12
Sub_parts	(Rungs Support)
Draw	Points to a procedure that draws the ladder on the screen

Figure 5.7. The class Ladder showing some of the values and some methods (adapted from Morjaria, 1989).

SLOT/ATTRIBUTE	VALUE
Name	Ladder-A
Type	Extension
Height	[120in]
Width	[18in]
Depth	[2.4in]
Rung_count	[10]
Sub_parts	(Rungs-A Support-A)
Draw	Points to a procedure that draws the ladder on the screen

Figure 5.8. The instance Ladder-A of the class Ladder (adapted from Morjaria, 1989).

limited range of choices is allowed. Procedures are given for calculating some ladder parameters, such as the width and the number of rungs, and there is a pointer to a procedure for drawing a picture of the finished ladder. The different types of ladders (Sub_classes) subsumed by the class Ladder also are identified.

The identification of the ladder's Sub_classes and the slot Sub_parts point to an underlying structure and to linkages that make the object data structures outlined in Figures 5.1 and 5.7 quite rich and powerful as descriptors. Figure 5.8 shows the *instance* of the class Ladder – that is, a particular example of one of the members of that class, Ladder-A. The instance is identified through a link that states, for example, that Ladder-A *is_a* Ladder. This particular ladder is a 10-foot-long extension ladder that has the specified parameter values and parts shown. The attributes are inherited from the class Ladder through the *is_a* link, with the values specific to Ladder-A being calculated in accordance with the procedures identified (and maintained) in the class description (Figure 5.7).♦ The pieces used to make the ladder are identified by still other links; for example, we could say that Ladder-A *has_part* Rungs-A. The details of the machinery required to support this structure are not important here. Rather, we see that the structure enables us to describe objects, both conceptual and "real," and to describe classes of objects in a rich yet compact way. The particular attributes can be chosen to meet the needs of any particular design project, as we have seen with design-with features and will see in subsequent discussion.

Inheritance can quickly get complicated. For example, we need a strategy to decide which path to choose if more than one class higher in the path up the hierarchy provides a value for the same attribute. Often, we use the first one found (i.e., closer above) because that is a more specific description. If there is no strict hierarchy, as happens when an object belongs to two classes (e.g., "connector" and "molded part"), then inheritance could produce contradictory information. If properties can be "defeated" (e.g., if one particular bolt is not silver), then inheritance needs heuristic guidance. Fortunately, description logic uses only strict inheritance, although rules and restrictions can be added, albeit at a price of decreased efficiency. Earlier structured representations (e.g., KL-ONE) allowed such things as arbitrary restrictions on "slot" fillers, as well as the ability to specify relationships between the fillers of slots (i.e., between the values of attributes). See Brachman and Levesque (2004) for more details.

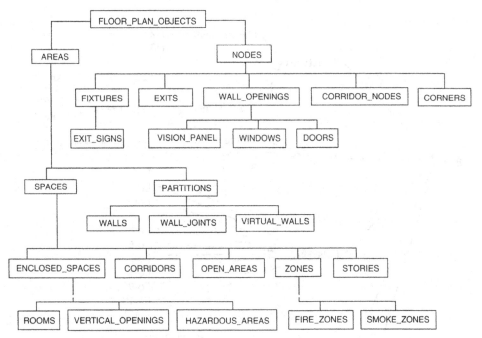

Figure 5.9. The object tree (or inheritance lattice) in the LSC Advisor KBES (Dym et al., 1988).

Figures 5.9 and 5.10 show two inheritance lattices or object trees of all of the objects in two KBESs.♦ The first is from the LSC Advisor, which was designed to check architectural plans for conformance with a fire safety code. The second is from the DEEP system for designing electrical service for lots in residential developments (cf. Section 5.1). The objects in the LSC Advisor are organized into a hierarchy of subclasses of the class FLOOR_PLAN_OBJECTS. Note that in addition to physically realizable objects, this hierarchy explicitly details conceptual objects that are crucial to the architectural design process (e.g., fire zones and smoke zones). Furthermore, in this representation, *all* the objects have in common the slot

> Figures 5.9, 5.10, and 5.11 are examples of the basics of ontologies. An *ontology* defines the kinds of objects, their properties, and relationships between them that a person or system knows about, or wants to deal with, for a particular task. Ontologies also define, as a side effect, a vocabulary of terminology. Noy and McGuinness (2001) provide a simple introduction to creating any ontology, and Ahmed et al. (2007) deal specifically with engineering ontologies. Gómez-Pérez et al. (2004) provide a comprehensive, general overview of the field with examples.

LSC_PROBLEM, which is defined in the class FLOOR_PLAN_OBJECTS. Some of the objects – also called *frames* – from this lattice are shown in Figures 5.16 and 5.17. We discuss these frames and the LSC Advisor in more detail in Section 5.5.

Figure 5.10 shows the inheritance lattice of all of the objects in the DEEP system that is used to configure the electrical service equipment for lots in residential developments. We previously displayed (cf. Figure 5.1) a particular instance (XFR50_1) of

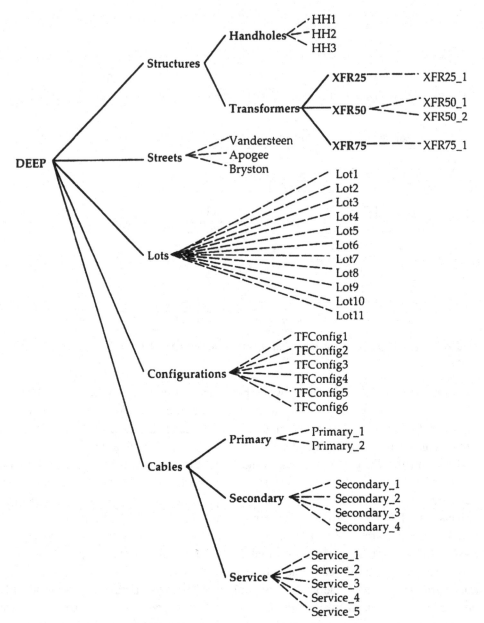

Figure 5.10. The object tree (or inheritance lattice) in the DEEP KBES (Demel et al., 1992).

the class Transformers, and we see here how that class and its instances fit into the overall structure of all of the objects involved in this particular design project. In fact, the classes Transformers and Handholes (which are typically small underground structures containing electrical connection and relay devices) are themselves subordinate to a more abstract class or "superclass," Structures, which is in turn part of the top-level class that contains all of the objects in the DEEP knowledge base. Notice the heterogeneity of this lattice; that is, it represents a mix of different kinds of physical structures and of more conceptual objects, such as the Configurations, which represent recommendations for the spatial layout of the cables and other electrical equipment.

What also may be evident from the discussion, although probably not from the figure, is that the branches of the object tree are not independent of one another, even though they look that way in the pictorial representation of the tree. For example, the transformer instance XFR50_1 has associated with it a specific location (i.e., the slot Coordinates in Figure 5.1) and a set of lots that it serves (not shown in that figure, although the total number of lots served is given in the slot MaxCustomers). The location of this particular transformer and the lots it serves are identified in appropriate slots in objects elsewhere in the tree; that is, in the classes Configurations and Lots, respectively.

At this point in time, many KBESs exemplify the use of object-oriented techniques to describe objects as they are being designed. Indeed, we could fill the remainder of this chapter – and book – with examples from the recent literature. However, for now we content ourselves with introducing just one more example, from the domain of structural design, which also illustrates the current state of the art in terms of routine computer implementation and points the way to the discussion (in Section 5.5) on integrating representations. The example is derived from a KBES that builds descriptive models of simple engineering structures by interpreting their graphical descriptions. In a style perhaps akin to feature extraction (cf. Section 5.3), the KBES identifies some particular features of physical objects and categorizes them into classes of structural types. Figure 5.11 shows the hierarchy of structural objects and types contained in the knowledge base of that system.

Each of the structural objects depicted in the hierarchy of Figure 5.11 is represented by a frame (or object) in the object-oriented language. A view of part of the frame for the class of structural object Truss is shown in Figure 5.12. It looks very much like the class descriptions we have already seen, but it is worth noting that some of the attribute values shown are in fact labels for procedures that are used to evaluate certain properties of a truss. Thus, for example, truss_span, truss_triangulation, truss_redundants, and truss_stability are the names of procedures or algorithms that are used to evaluate any instance of the class Truss to assess whether that instance does have the requisite properties of a structural truss.

We should remember that our partial view and description of this frame leaves some things undescribed. In particular, although we have specifically identified by name the procedures that are called to obtain the values for the last four attributes, we have not distinguished them in any obvious way from the ways by which other slots are assigned their values. As with previous examples, some values are inherited from higher classes (e.g., line_element_structure), some by virtue of being in a certain class (e.g., NumberOfJoints), and so on. As we remarked in the Preface, there is much variation in style of the pseudo-code that can be used to illustrate objects. Thus, although we do mimic the styles of the generators of each object, we will not go into all the detail needed to make explicit the particular meaning of every facet of each format.

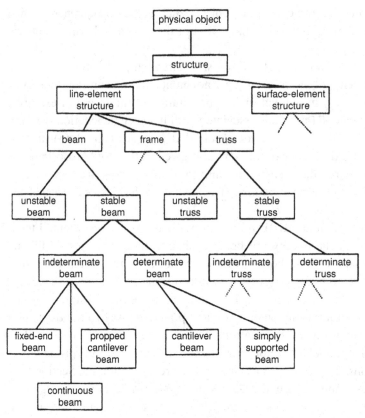

Figure 5.11. The hierarchy of structural objects in a KBES that interprets structural drawings (Balachandran and Gero, 1988).

SLOT/ATTRIBUTE	VALUE
SuperClasses	line_element_structure
SubClasses	[stable_truss unstable_truss]
MembershipCriteria	[numberOfMembers numberOfJoints numberOfSupports typeOfJoints typeOfForces typeOfSupports locationOfSupports]
NumberOfJoints	[intersection integers greater_than (2)]
NumberOfMembers	[intersection integers greater_than (2)]
NumberOfSupports	[intersection integers greater_than (1)]
LocationOfSupports	joint_locations
Span	truss_span
Triangulation	truss_triangulation
NumberOfRedundants	truss_redundants
Stability	truss_stability

Figure 5.12. The class Truss in the structural-drawing interpreter (Balachandran and Gero, 1988).

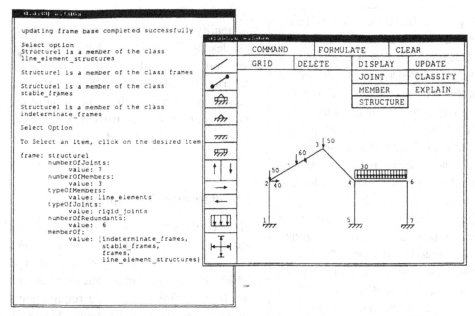

Figure 5.13. The interpretation of a two-bay frame (Balachandran and Gero, 1988).

Figure 5.13 shows the first of two sample runs of the interpretation system. We note that the system has identified a graphical description of a gabled two-bay frame, recognizing that it is stable and indeterminate to the sixth degree.

Figure 5.14 shows that the apparently unnamed system – which is itself an interesting phenomenon because it often seems that many KBES designers agonize over

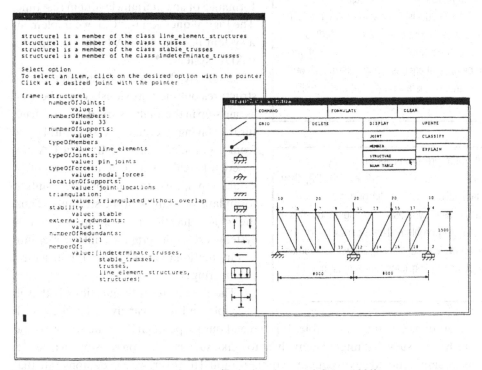

Figure 5.14. The interpretation of a truss (Balachandran and Gero, 1988).

Advising systems can be formed in a variety of ways. Critiquing systems (Oh et al. 2008; Silverman 1992) work by taking users' decisions (e.g., a complete or partial design) and providing critical feedback that may cast doubt about some aspect. Critics can comment on a range of things, from actual errors (e.g., geometry that cannot be manufactured) to potential life-cycle problems (e.g., a particular kind of connection is prone to failure in use). A particular criticism can be formed by comparing a user's decisions against those already made, or retrieved, by the advising system. Alternatively, we can store patterns known to often cause problems, or be suboptimal, in the advising system and then match them against the user's decisions – which is why critics are often rule-based. Critiquing systems also can be used in a "proactive" manner, providing suggestions and guidelines before the designer acts (i.e., critiquing the situation). For example, some case-based reasoning systems provide a case base of past problems that have occurred during designing, pointing to potential difficulties with certain requirements or decisions that should be avoided. Constraint-based systems (O'Sullivan 2006) also can be used to check design decisions as they are made (to see whether any constraints are violated) or, for example, to check the specifications of configurations problems that are described in terms of conditional constraint satisfaction (Finkel and O'Sullivan 2011).

Although there are examples of integrated representations in systems (in academia mostly), they are usually tailored to specific purposes, such as functionally based synthesis or configuration. It is a very challenging problem. There certainly are systems that provide access to and allow management of product or building information (Eastman et al. 2011).

the choice of a suitable acronym with which to identify their work – identified the graphical description of a truss that has 33 members and 18 pin joints, and that it recognizes the truss as being indeterminate to the first degree. We also note the structuring of the information about the truss in the pattern defined in its frame-based description (Figure 5.12).

5.5 Integrated Descriptions

The descriptions outlined in the previous section raise some interesting questions. For example, the LSC Advisor, a prototype KBES, checks that an architectural design conforms to fire-safety–code requirements on such aspects of a building's floor plan as the extent of fire and smoke zones, the location of exits and exit signs, and egress.♦ Thus, the LSC Advisor would be most helpful to an architect if it were able to sit astride the CADD system in which a building floor plan was being designed and check the conformance of an emerging layout in real time. The logical question is: "Can we implement a system like the LSC Advisor on top of a CADD system?"♦

Similarly, the graphical interpretation of structures outlined previously could be seen as one step in the analysis and design of structures. Having identified a structural type from a graphic, which might be a CADD-based "sketch" of a structural concept, it would be helpful if we could use the results to formulate an analysis of the structure. Thus, the logical question is: "Can we integrate a system such as the graphical interpreter into a computational environment for structural engineering?"

The answer to both questions is that in principle and increasingly in practice, such integrations are indeed possible. It is beyond our scope to analyze the software and hardware issues that might be involved to make such integration possible in a particular computational environment or organization. However, we can demonstrate that the merging of the underlying representations is quite logical and, in principle, rather straightforward. We do this by exhibiting some aspects of the representations used in the LSC Advisor, including rules, objects, and a direct connection to the graphical representation of a floor plan in a commercial CADD system called GraphNet™. Our account of how that KBES works, however, is very terse. The rule and frames we discuss are shown in Figures 5.15–5.17. The object tree into which the frames (or objects) fit is shown in Figure 5.9.

The floor-plan representation developed in the LSC Advisor was designed to make it easy to describe the objects in a typical blueprint and to reason about these objects in the context of analyzing and applying the particular fire code applied, the *Life Safety Code* (LSC) of the National Fire Protection Association. As already observed, the objects in a typical floor plan (e.g., fire zones, rooms, doors, and walls) are fairly abstract, so rules alone are incapable of describing such objects, especially when they have many complex attributes. Thus, the combination of rules and frames represents a first level of integration.

The rule in Figure 5.15 describes properties (WALL_TYPE) that particular walls (?WALL) forming the boundary (?ZONE) of a fire zone (FIRE_ZONES) must have in order to comply with the LSC. The identification of particular physical and abstract objects is done within the CADD system – which serves as the front end to the LSC Advisor – by the architect. He or she must explicitly identify the floor plan's different elements (e.g., the rooms, doors, and walls) and select room labels from a special list so that the occupancy type (e.g., office and patient's room) can be properly determined. This allows direct and automatic translation of the CADD database into the frame-based floor-plan representation that the LSC Advisor analyzes. As shown in Figure 5.15, the antecedents of the rules representing the LSC requirements (i.e., the rules' left-hand sides) contain the clauses necessary to trigger their application.

The inheritance lattice for the LSC Advisor, shown in Figure 5.9, points to the handling of objects in groups. All objects of the same type are represented as instances of a class; for example, all rooms are instances of the class ROOMS, all doors are instances of the class DOORS, and so on. Two instances from the LSC Advisor's hierarchy are shown in Figures 5.16 and 5.17.

```
IF:     (AND   (?ZONE IS IN CLASS FIRE_ZONES)
               (THE REQUIRED_ENCLOSURE_RATING OF ?ZONE IS 2)
               (THE BOUNDARY OF ?ZONE IS ?WALL)
               (THE WALL_TYPE OF ?WALL IS INTERIOR)
               (AN OPENING OF ?WALL IS ?OPENING)
               (LISP (< (THE RATING OF ?OPENING) 1.5)))

THEN:   (AND   (A LSC_PROBLEM OF ?ZONE IS
               "FIRE RATING OF WALL OPENING TOO LOW:
               ?OPENING")
               (A LSC_PROBLEM OF ?OPENING IS
               "FIRE RATING SHOULD BE 1.5 HOURS.
               ACTUAL RATING: ?RATING"))
```

Figure 5.15. A rule in the LSC Advisor (Dym et al., 1988).

SLOT/ATTRIBUTE	VALUE
CORNER	(# [Unit : CRNR99 BUILDINGS]
	# [Unit : CRNR104 BUILDINGS]
	# [Unit : CRNR94 BUILDINGS]
	# [Unit : CRNR32 BUILDINGS])
EXTERIOR_NODES	(# [Unit : CRNR97 BUILDINGS]
	# [Unit : CRNR102 BUILDINGS]
	# [Unit : CRNR92 BUILDINGS]
	# [Unit : CRNR37 BUILDINGS])
GROSS_AREA	(2408.0)
OCCUPANT_LOAD	(6200.0)
SUBSPACES	(ROOM180 ROOM151 ROOM152)
ZONE_CORRIDORS	(CORRIDOR1 CORRIDOR2)
ZONE_DOORS	(DOOR79 DOOR86 DOOR89 . . .)
ZONE_EXITS	(DOOR79 DOOR86)
ZONE_EXIT_SIGNS	(EXIT_SIGN1 EXIT_SIGN3 . . .)
REQ'D_ENCLOSURE_RATING	1.5
NET_AREA	(1956.0)
LSC_PROBLEM	NIL

Figure 5.16. An instance of the class FIRE_ZONES in the LSC Advisor (Dym et al., 1988).

We see in these two instances how the floor-plan–representation hierarchy explicitly relates geometrical information to high-level objects such as walls and doors, requiring that the architectural CADD system provide such a unified description of the building. The class NODES, one of the immediate subclasses of FLOOR_PLAN_OBJECTS, defines a COORDS slot for storing a single (x, y) location. One of the subclasses of NODES is CORNERS, which specifies the perimeter corners for any instance of FIRE_ZONES. The instance also has a slot for the set of EXTERIOR_NODES it has on the building's exterior walls. The instances also have slots for other attributes that are important for conformance checking, such as the GROSS_AREA, OCCUPANT_LOAD, ZONE_DOORS, and ZONE_EXIT_SIGNS. We also can store detailed information on construction materials, wall finishes, and product names for prefabricated items in the slots' parent classes (e.g., DOORS and EXIT_SIGNS).

The floor-plan information represented here goes well beyond the purely geometric data of a blueprint. By unifying these diverse types of information, we take a step beyond most architectural CADD systems, which do not have unified databases

SLOT/ATTRIBUTE	VALUE
CORNER	(# [Unit : CRNR113 BUILDINGS]
	# [Unit : CRNR112 BUILDINGS]
	# [Unit : CRNR139 BUILDINGS]
	# [Unit : CRNR138 BUILDINGS])
WALL_TYPE	INTERIOR
RATING	2.0
LSC_PROBLEM	NIL
WALL_OPENING	(DOOR153)

Figure 5.17. An instance of the class WALLS in the LSC Advisor (Dym et al., 1988).

for materials and floor plans. Floor plans often are generated by drawing programs with separate databases of lines, arcs, and text strings. Such graphical databases contain no indication, for example, as to which lines should be understood as representing the two sides of the same doorway. This information can be deduced only by the architect from a complete drawing, unless just such an integration of the representations is available.

We can also use these two instances to show how the combination of rules and objects in the LSC Advisor enables efficient evaluation of the fire-resistance rating of a door in a wall intended to serve as a fire barrier. How can we rapidly and efficiently identify specific doors in a floor plan that may contain several thousand objects? First, we can identify all instances of fire zones that are direct descendants of the class FIRE_ZONES, such as shown in Figure 5.16. This list of objects could have between two and five entries. Because the class FIRE_ZONES is a subclass of the class SPACES, it has a slot for the attribute BOUNDARY, which lists all of the walls that comprise the boundary of that fire zone. This list of walls, with anywhere from 4 to 30 entries, is winnowed down to those with the value INTERIOR in their WALL_TYPE slot. Members of the class WALLS (e.g., Figure 5.17) contain the slot WALL_OPENING, which lists all the DOORS in a given wall, so we now have identified the specific doors to which the relevant rule applies. We can examine the RATING slot of each door so identified and apply the relevant rule.

There are other interesting aspects of the LSC Advisor's representation scheme, including the integration of some complex geometrical calculations and algorithms with the object-oriented representation. The most interesting (and complex) of these algorithms are those that measure *travel distances* to the building's exits from various points in the building. These algorithms are needed because the LSC sets upper limits on the travel distances from room doors to the nearest exit. The travel-distance algorithms perform the distance-measuring calculations with data derived from the graphical information; they use the object representations and mechanisms to pass the calculated distances to the appropriate slots where they can be held until it is time to apply the appropriate rule. As mentioned previously, however, this is not the place to review all of the LSC Advisor's facets. We mention these other features only to reinforce the idea that integrated representations of designed objects can be implemented to produce powerful results.

From the viewpoint of design as a discipline, there are obviously some very good reasons to work toward the integration of representation. A full-blown implementation of an integrated system combining a KBES such as the LSC Advisor with an architecturally oriented CADD system and a database of typical architectural components obviously could enhance the architectural-design process. Similarly, structural engineers would find it useful to have an integration of representations that allowed them to select members, analyze them, and check them against design codes, such as the *Load and Resistance Factor Design (LRFD) Specification for Structural Steel Buildings* published by the American Institute of Steel Construction (AISC), as well as check them against national and local building codes, such as the previously mentioned LSC. To complete our discussion of

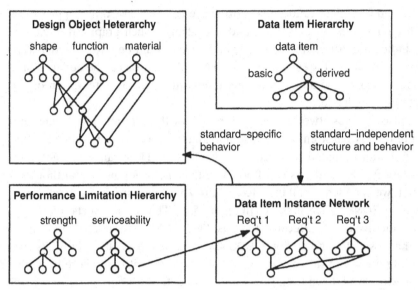

Figure 5.18. An object-oriented model of a design standard (Garrett and Hakim, 1992).

integrated representations, we show elements of an object-based description of the LRFD specifications in Figures 5.18–5.21. Two kinds of integration appear in these examples. In the first, procedural knowledge expressed in algorithmic form appears within the slots or attributes of some objects. In the second, paralleling to some extent the descriptions of the LSC Advisor and the graphical interpreter already described, concepts are integrated in ways that lay a foundation for further developments in representing designed objects.

Figure 5.18 shows how a design standard can be represented in an object-oriented way, using as an illustration the requirement that structural columns must meet to avoid flexural–torsional buckling. The top-level of this collection of objects is a *design object heterarchy* that does two things. It identifies the classes of components and systems to which the design standard applies, which in the present example are the sets of structural columns shown in the partial design object heterarchy depicted in Figure 5.19. Each element in Figure 5.19 is part of the parent class column that is shown in Figure 5.21. Note that a data slot, FTB, is created in this class to represent the requirement imposed by the standard. Then, an instance of a requirement rule is created as part of the definition of FTB, so that each and every column in the class column will inherit this rule and be forced to adhere to its meaning.

Figures 5.20 and 5.21 show details of the descriptions of columns in terms of their various attributes. And, to illustrate how object-oriented standard processing works, we outline a performance check of a particular wide-flange column, *W 14×34*, which is subject to a compressive force of 150.0 kips (or 150,000 lb). First, from the heterarchy (see Figure 5.19), we create an instance of this column, W–ds–column, to reflect the fact that we are checking a W-section, doubly symmetric, I-shape column. Using the technique of *message passing*, by which objects communicate with each

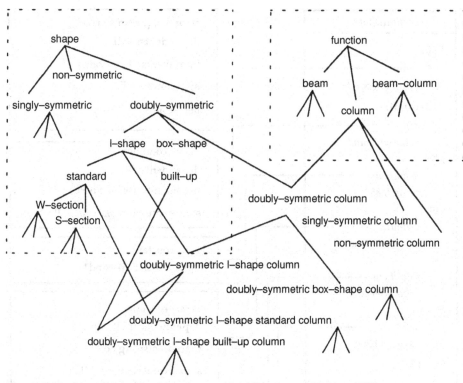

Figure 5.19. A partial design object heterarchy (Garrett and Hakim, 1992).

other in the implementation of object-oriented programming, the lack of an initial value for the slot FTB in our new instance is communicated to the object FTB, which in turn asks for a value for the slot FTB-cond1 to determine whether the condition has already been satisfied. Because it has not, a chain of requests for data is set in motion because before the calculation can be made, we need to know the values

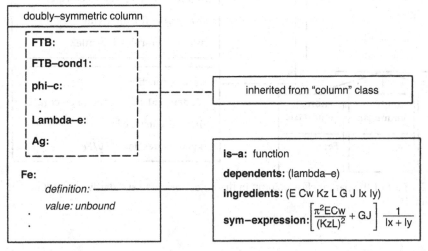

Figure 5.20. Part of the class doubly-symmetric column (Garrett and Hakim, 1992).

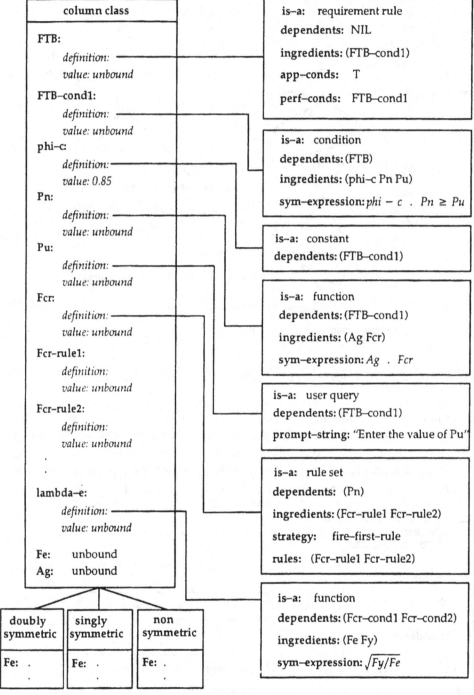

Figure 5.21. Part of the parent class column in the object-oriented standards model (Garrett and Hakim, 1992).

of parameters such as ϕ_c (phi–c), the ultimate and nominal resistances (Pu, Pn), the gross area (Ag), and critical stresses (Fcr, Fe, Fy). In some cases, the user (i.e., the designer) is prompted for the information; in some cases it is found by looking up values in a referenced table; and in other cases, it is computed from formulas already integrated into the object descriptions (Figure 5.20). When we finally complete the calculation for this case, we find that the definition of FTB–cond1 can compute its value as true (T) and insert it in the app–conds slot of FTB because the subject column does satisfy the relevant design standard.

We point out again that this brief discussion of the use of an object-oriented language to describe the application of a design standard (or the checking of a design for conformance with a code) is intended to show the integration of different kinds of knowledge that are expressed in different forms – for example, as values to be looked up in tables or as results to be obtained from the application of specified formulas. Thus, the notion of integration here is more in terms of types of knowledge and concepts than just of language or representation – and a formal taxonomy of engineering knowledge types has only recently been proposed – but it clearly is still in the mold of the kind of integrated representation of designed artifacts that we are discussing. However, it does set the stage for the final section of this chapter, wherein we discuss issues of integration in the context of communicating about designed artifacts.◆

An editorial comment may be appropriate before we leave this discussion of integrated representations. The reader will have noticed that much of this discussion seems at least implicitly aimed at the notion of integrating representations in a computational sense – and this after we have stated repeatedly (e.g., in the Preface and in Chapter 4)

> Aurisicchio et al. (2006) studied expert communication – in particular, information requests from expert designers. An example of an information request from one designer to another is "How can I retain the seal in place?" They were able to classify requests using several major categories, including the objective behind the request (e.g., trying to generate an explanation), whether it is about product or process, the type of response required (e.g., yes/no or an identifier of an information source), and the process needed to respond to the request. Other aspects included the kind of reasoning that was the cause of the request (e.g., diagnosis or functional decomposition) and the things mentioned in the request (e.g., predicted, observed, or intended behavior). Note that this analysis relies heavily on a structure-behavior-function model as well as knowledge of the types of reasoning present during designing.

that we are interested less in rendering design computable than we are in understanding how we think about designed objects and design processes! We maintain, however, that we are not being as inconsistent as it might appear at first blush. As humans, we are used to thinking about objects and artifacts in an integrated way, using different representations as appropriate to answer different questions, perform different tasks, and solve different problems. So, here we view the computer as, loosely speaking, something of a metaphor for modeling how we think–that is, in terms of different abstractions based on different representations which is a thought quite consistent with the quotation from Knuth (cf. p. vii) that was one of the starting points of our journey.

5.6 Communicating about Designed Artifacts

There has long been a gap between the designers of artifacts and their makers. Whether a natural evolution of the engineering profession or a reflection of Adam Smith's model of specialized labor as a driving force in capitalism (although it is also evident in societies built on vastly different economic premises), this gap has nonetheless become a source of significant concern in today's very competitive global economy. One of its most common expressions is the desire to tear down the legendary "brick wall" that is said to exist between designers and manufacturing engineers. The standards processing system just described can be taken a step further to address such concerns by generalizing the integrated representation of an object to a *product model* – that is, a database file within a design organization that contains all of the information (and its relevant pointers) generated about a specific project.

Figure 5.22 shows one scheme for structuring a product model for architectural design; it includes both a CADD representation and code checking or standards processing. This product-model description could serve several purposes. With regard to code checking, an integrated design environment would provide the context for identifying which standards or code provisions to apply. In the column evaluation just described, for example, because the chosen wide-flange section could be used as a beam or as a column, some way must be found to direct the processing to those standards pertinent to column design. An integrated system would also facilitate the gathering of the data needed to determine whether the relevant standard has been met.

The example underlying the illustration in Figure 5.22 is the evaluation of an object called a building to see whether it conforms to the standards delineated in the *Uniform Building Code (UBC)*. Per the foregoing discussion, we first identify which parts of the UBC apply to the particular building specified in Instance–232. In the example, based on the data provided in the product model, the standards processing system has found that Instance–232 is covered under the UBC as a Group–A–Division–2–Type–II–F.R. type of building. The standards processor also creates an instance of the context from the relevant portion of the code, Instance–102, and links it to the product model, Instance–232. Then, the data defined in Instance–232 are used to evaluate the two requirements in Instance–102 of type Group–A–Division–2–Type–II–F.R.

This simple example contains within it the seeds of several interesting issues about design communication. One of these is related to understanding the technical aspects of integrating CADD and other representations (e.g., object or device representations, analysis programs, codes, documentation, and so on) within an integrated computational design environment. A second issue – or, more accurately, set of technical and institutional issues – is concerned with the concurrent use of design software by many users in large design and manufacturing organizations. For example, in addition to the technical issue of integrating different representations, we have to deal with the need for different kinds of information.

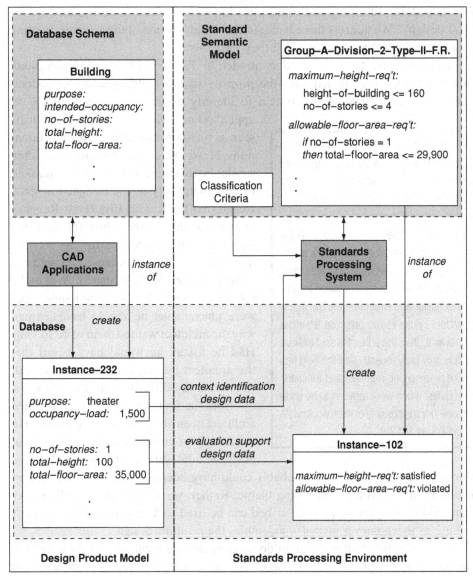

Figure 5.22. Interfacing CADD systems with standards processing and design product models (Hakim and Garrett, 1992).

We also must confront the issues of maintaining control of an evolving design, understanding how and why design decisions are made, and understanding how these decisions are propagated and enforced.♦ Wanting to understand *why* a design decision was made – an issue also raised in the previous example – prompts us to ask the question: "Do we gain some advantage

Engineering change (Jarratt et al. 2011) refers to changes made to a design during or after the design process. Such changes can cause other changes, which propagate through the design. If we know the reasons behind design decisions (i.e., the rationale), then we can better manage risk as we propagate design changes.

by articulating, preserving, and communicating the intentions that designers have for their design?" We address these issues in this (final) section of this chapter, treating them inversely to the order in which they were raised.

We note first that the product model in Figure 5.22 has communicated, in an appropriately labeled slot, the *purpose* of the building under consideration, which here enabled the system to identify which section of the UBC to apply.[♦] As noted in Chapter 2, designs can be seen as realizations of their designer's intentions. However, design intentions are often subtle in their expression or are masked by the complexity of the designed artifact. Recall that the Kansas City Hyatt Regency failure was due to the fabricator changing the structural connections for the second-floor atrium walkway because he could not hang it as originally designed. When the fabricator found that the long hanger rods were unavailable, he should have learned why the architect wanted them to be so long. Had he asked, he would have heard that the architect intended to hang the second-floor walkway directly from the roof truss, not from the fourth-floor walkway. Had the architect been able to signal this intention to the fabricator automatically and unambiguously, without waiting (in vain) for the question to be asked, this tragedy probably could have been avoided. But, our interest is not in finding fault or attributing blame. Rather, we wonder whether the representation techniques we have described can be used to help convey the designer's intentions to whomever makes or assembles the resulting design.

> "Purpose" plays a role in design rationale and in functional representations. Designs are triggered directly or indirectly by a *need* (i.e., something desired) – for example, to get to another place. Users' – and perhaps designers' – *goals* (e.g., desired states) are triggered by the need, with the intention to satisfy that need (i.e., a *purpose*). Intentions are sometimes seen as very abstract plans. Functions are an intermediate level between an artifact (i.e., its behavior and structure) and the intention. If an artifact in use allows progress toward a goal, or achieves it, then the artifact is said to have a function. It is only fairly recently that the field has thought about representing rationale and functions in this much detail: much work remains to be done to establish whether it is possible or even worthwhile (Vermaas and Eckert 2013).

Capturing the designer's intent does not seem so difficult, especially in light of the building example we just saw. Surely, we could attach one or more slots to each and every object description that would allow us to call out the reasons we are doing something with that object. For example, we could imagine an object description of the second-floor walkway something like that depicted in Figure 5.23. Using made-up names for both slots and their values, we can clearly describe a hanger support system for a walkway in a form that leaves no doubt about the ultimate source of support of the walkway, or that the path the load must travel is directly to the roof truss, or that there are no intermediate connections. Conjuring up a description like this is fairly easy. The real trick is to get this information into the hands of the fabricators, the people who will actually build what we design.

We should note, however, that this view of capturing design intent will seem elementary or superficial to some because it leaves the solution to the problem of transmitting that intent at the same abstract level as our discussion of integration. Furthermore, it ignores some recent attempts to use design intent as a point of departure for reasoning about design. To choose a less dramatic example than the Hyatt Regency, consider the following circumstance. A contractor is looking at the specifications for a building – including both the drawings ("plans") and the documentation – as she prepares to choose a fabricator for the structural frame as part of her bid submission. She notes that a particular steel is called out for the frame members (say, A242) and wonders if a less expensive steel could be substituted. If the design documentation specified the history of the design process as well as its results, the contractor would learn that A242 steel was specified for its ability to resist corrosion, and that this choice was predicated in turn on the fact that the architect's design called for the steel frame to be exposed so that he could highlight the structural function as part of his aesthetic for the building. (This aesthetic approach has, by the way, led to some spectacular buildings, including Pittsburgh's US Steel building and Chicago's John Hancock tower.)

Now, we note in Chapter 3 that the design process can be viewed as one of refining abstract goals and objectives – that is, rendering them in increasing detail until a fabrication specification emerges at the end of the process. In this context, *design intent* can be viewed as a recorded history of the design process in which the *reasons* that design decisions were made (or that particular refinements were generated) are tracked as they are implemented. That is, if we could capture design intent in a systematic fashion, we could review design decisions and understand the consequences of revising them or undoing them altogether. More than simply a retrospective glance at the process, design intent becomes a reasoning tool that can help guide verification of a design, modification of the design as changes are proposed, or reuse of the design for some other purpose.♦ With reference to our made-up example (cf. Figure 5.23), we could perhaps imagine a KBES or other reasoning system that has a set of objects that describe load paths in the proposed building design. Then, this system could be queried about connections to which loads could be transferred or about alternate designs (or consequences) if we wished to modify existing connections or insert new ones.

> Recent work on design rationale (DR) (Burge and Bracewell 2008; Burge et al. 2008) provides most of these capabilities. Burge's work allows both the syntax and the semantics of the DR to be checked automatically. For example, selecting a design decision that was not as well supported as another, or that was missing some rationale, would be flagged by the system as suspect. Bracewell's DR research (i.e., the DRed system) provides comprehensive graphical capabilities for displaying large amounts of rationale.

Figure 5.24 shows two instances that we made up to illustrate just how the different load paths might appear to the user of such a system: the top instance shows that the second-floor walkway is connected directly to the roof truss and carries no other load; the bottom instance reflects what the Hyatt Regency fabricator did (that he should not have!). In addition to the reasoning potential for design, it clearly would be quite valuable to make building plans – including the architect's

SLOT/ATTRIBUTE	VALUE
TYPE_OF_HANGER	THREADED_ROD
HANGER_CONNECTION	NUT_ON_WASHER
NO_OF_HANGER_PAIRS	4
LOCATION_OF_HANGERS	(# [2nd_WALK_DIST030]
	# [2nd_WALK_DIST060]
	# [2nd_WALK_DIST090]
	# [2nd_WALK_DIST120])
CONNECTS	(2nd_WALK ROOF_TRUSS)
INTERMEDIATE_CONNECTS	NONE
LOAD_TRANSFER_TO	ROOF_TRUSS
LOAD_TYPE	TENSION

Figure 5.23. An object description for the HANGER for the second-floor walkway for the Kansas City Hyatt Regency Hotel.

and structural engineer's intentions – available to the contractor and the fabricator in this way.

Figure 5.24 shows that through the creation of the slot SUPPORTS, we identified the *functional* purpose of the hangers to which they are linked (by virtue of values in the LOAD_CARRIER slots in Figure 5.24 and the presumed links to the HANGER class shown in Figure 5.23). America's most noted architect, Frank Lloyd Wright, is also famous for the dictum that "Form follows function." The representation of function and its links to form and its representations is also a lively area of debate. Does the specification of function dictate a specification of form? Can we infer function from form?

We can easily argue from the examples seen in this and the preceding section that the answers to both questions are, in general, negative. Certainly, the specification of a structural element to carry compressive loads between floors dictates that we

SLOT/ATTRIBUTE	VALUE
LOAD_PATH	(2nd_WALK ROOF_TRUSS)
SUPPORTS	2nd_WALK
INTERMEDIATE_CONNECTS	NONE
OTHER_LOADS	NONE
LOAD_TYPE	TENSION
LOAD_TRANSFER_TO	ROOF_TRUSS
LOAD_CARRIER	TYPE_OF_HANGER
LOAD_PATH	(2nd_WALK ROOF_TRUSS)
SUPPORTS	(2nd_WALK 4th_WALK)
INTERMEDIATE_CONNECTS	4th_WALK
OTHER_LOADS	(4th_WALK_LIVE 4th_WALK_DEAD)
LOAD_TYPE	TENSION
LOAD_TRANSFER_TO	ROOF_TRUSS
LOAD_CARRIER	TYPE_OF_HANGER

Figure 5.24. Object descriptions for two instances of the class LOAD_PATH showing different load configurations.

need columns, but the shape of any particular set of columns is contingent on many more factors than the simple need to carry a compressive force. Similarly, as we commented about the wide-flange section in the discussion of standards processing, we could not infer from its form how it was to be used because such I-shaped elements can be used as both columns and beams or girders. We could make such an inference if we knew more about the context in which we were designing. For example, if we were using wide-flange sections for both columns and girders in a rigid frame, we could look at their relative stiffnesses and infer function from form by noting that column stiffnesses are designed to be larger than girder stiffnesses by a factor of 2.5 so as to guarantee stability of the frame (by ensuring that any plastic hinges that form occur first in the girders rather than in the columns). This brings us back to the discussion of intent, both as a guide to design and as data that need to be transmitted to whomever is in charge of making the designed artifact. As an expression of design philosophy and reasoning, the foregoing column–girder rule of thumb represents a refinement of a higher-level goal of avoiding frame instability in the presence of seismic loading.

We thus see the need to interpret, refine, and represent at different levels of abstraction the intentions we seek to realize in a design. The question is: "Are issues of intention for and function of an artifact inextricably linked to the issue of representing the artifact?" That is, can we think of function independently of the representation of the device that performs that function?♦ We can see from the examples, in fact, that the details of the representation of the artifact can be chosen apart from the representation of that function, subject only to the proviso that in the final analysis we, as designers, have to produce a set of complete and unambiguous specifications for the manufacture of that artifact, which means that this representation must be acceptable to the manufacturer.♦

We return now to the issue of communicating about the design, both as it unfolds and as it must be finally conveyed to the manufacturer. As we have said before, whereas

The independence of function and structure was discussed previously. However, a related issue is how much of the context of a device is included when describing its function. Chandrasekaran and Josephson (2000) point out that much of the description of function in engineering is "device-centric" (i.e., describing its properties or behavior with no reference to what effects it has on the environment in which it is located). An example is the classic Pahl and Beitz view of defining a device by its input/output of energy, material, and signals. Because engineered devices always have a particular desired effect (i.e., an intended function) that is usually well understood by all parties, it is often not explicitly mentioned. For example, "a pen produces ink at its tip" does not mention the intended function. However, a complete description of function requires an "environment-centric" description (e.g., a pen produces ink at its tip so that with proper placement and movement against some medium, ink can flow out to make useful marks).

It is interesting that functional modeling work usually makes no reference to manufacturing, although that is understandable given that it is not usually concerned with detailed design. Erden et al. (2008) provide the most comprehensive review of function modeling, whereas Hirtz et al. (2002) describe the product-specific, engineering design view (i.e., the functional basis).

the issues are both technical and organizational, we think that they are not so much about representation principles or about the pragmatics of software and hardware. We address the organizational issues first, starting with the fact that the design of most complex artifacts requires combining the expertise of specialists in several discrete areas. The various kinds of expertise can be deployed either sequentially or concurrently. In sequential design, we break down the design task into a sequence of subtasks for which the output of one designer's work serves as input to the next. In concurrent design, specialists in different disciplines (e.g., structural, mechanical, electrical) or functions (e.g., planners, accountants, operators) work simultaneously – but separately. In this case, it is of paramount importance that we coordinate design decisions in real time or periodically review them for consistency and modify them as needed during periodic reviews. Furthermore, in this case, which is crucial for the design of complex or substantially innovative artifacts, we need to understand that the experts or designers working on a piece of the complex artifact will work in the representations they prefer for their task and on the current state of the design as they know it.

We are still evolving models of concurrent design and for its support. Furthermore, because we can argue that modeling such design is modeling the design process, we defer until the end of the next chapter the description of some architectures that have been proposed for concurrent design. For now, it seems reasonable just to glance at concurrent design with an eye toward identifying the various artifact-representation issues that might arise.

As the conceptual design of a complex artifact unfolds, high-level specialists from all relevant design disciplines must (1) agree on the key performance specifications and constraints arising from each discipline, and (2) articulate and communicate the assumptions of designers from the participant disciplines because they could have implications for all other disciplines. As a conceptual design becomes firm, we can identify both key subsystems that fall primarily within the area of expertise of a single design specialist and the top-level interface issues that must be coordinated with other specialists. After we reach this stage, individual design activities can proceed concurrently, in a much more loosely coupled fashion. Individual designers develop partial design solutions covering the components, materials, dimensions, and other attributes of the subsystems assigned to them, and they communicate across discipline or subsystem boundaries only infrequently.◆

> Visser (2006) notes that her experiments show that design teams add the need for new types of knowledge to support collaboration, to allow them to manage inter-task dependencies, generate and understand external representations, and handle various kinds of communication (Kleinsmann and Maier 2013).

As the final, detailed descriptions of all subsystems are developed for a complex undertaking such as an aircraft or power plant, we are then faced with the overwhelming task of sharing and coordinating an enormous amount of information about a very large number of objects and devices because we have to ensure that all of these pieces fit – functionally, spatially, and temporally. In traditional design practice, this often involves the circulation to all affected parties of paper drawings that are "redlined" to identify conflicts, errors, or omissions. This process is serial in nature and thus extremely time-consuming. Moreover, it is certainly susceptible to errors.

However, as designers in more and more fields develop syntheses on their own workstations, their partially developed designs become machine-readable (and, as we mentioned herein, machine-translatable and interchangeable). That is, three-dimensional (3-D) CADD models provide a ready alternative to redlining for consistency checking and communication across disciplines. One version of an integrated 3-D model would have separate layers for each discipline's input, which could be easily used, for example, to check for *spatial consistency* ("interference checking"). It could eliminate the need for physical modeling of artifacts, such as scale models of refineries and full-scale mock-ups of new aircraft. In historical terms, CAD and CADD systems have been used largely for *documenting* design; only recently have they been used for spatial coordination of designs.

It must be said, however, that current computational design tools provide hardly any support for *automated design synthesis*, and they do little for design managers in verifying *functional consistency* among an artifact's subsystems. This indicates a need for not only the kinds of representational integration we have espoused but also for software architectures that address the need for intelligent functional coordination among subsystem designs developed by separate design teams. (We explore this issue in more detail in Chapter 6.)

Another obstacle to computer-aided design integration has been the incompatibility of database architectures for CADD and other systems used by different design specialists. We could force an administrative solution to this problem by insisting on the use of a single CADD package, assuming that it can support all of the kinds of analysis we need. A more evolutionary line of attack on this "Tower of Babel" design problem is the development of knowledge-based database interfaces that can mediate between several related but otherwise incompatible databases. Although we have said that software (and hardware) implementation issues are beyond our scope, it is worth noting the development of standards by which CADD systems could share their data (and representations) with other software. The vendors of CADD systems – perhaps strongly prompted by their customers – have for some time been interested in facilitating the exchange of data among different software packages. Most designers and their organizations do not want their options to be limited to a single piece of software or even to a single vendor; therefore, in response, professional societies and organizations have provided forums and means to standardize the format in which data are produced by CADD systems, thus enabling users to share results more easily. The first step toward this was the Initial Graphic Exchange Specification (IGES) standard, which was concerned solely with the exchange of drawing information, not interpretive information. Thus, for example, IGES was concerned with interchange between wire-frame representations and drawing packages. The

Product Data Exchange Specification (PDES/STEP) emerged more recently as an international standard for exchanging product information. Thus, the PDES/STEP standard is concerned with standardizing a format for many of the kinds of information we identified in the various examples and illustrations in this chapter. This is not to say that a standard for KBESs or object descriptions is emerging but rather that much of the product knowledge and information they will use and lodge within their own representations will be available in a standardized format, so the process of achieving the integration we have repeatedly stressed here will be substantially eased.

With regard to representational issues and integration, we hope that we have demonstrated the utility of representing (or describing) objects in different languages or styles, at different levels of detail or refinement, and in ways that invite an integrated view. We are, after all, simply trying to imitate what we do as human designers. We also should say that the discussion presented here has left many topics (and possibilities) untouched. One of the most important of these is the linking of artifact representations to analysis tools, whether they be expressions of formulas on paper, embedded in software for the symbolic manipulation of formulas, or exercised as numerical algorithms (e.g., finite differences, matrix methods, finite elements, and so on). Such integration is coming and its appearance is heralded by an increasing interest in using KBESs as front ends and user interfaces for complex analysis packages.

Commercial knowledge-based design systems that can be used to generate design syntheses for semicustom products are becoming increasingly available. These systems can be developed to run in a fully automated mode or they can keep human designers in the loop via human interface tools such as product-structure graphs and geometric CADD visualizations of the design, to which the human designer can react at each stage of design development. They are based on somewhat different styles of integration but, conceptually, they are very similar to what we have seen in the product model idea shown in Figure 5.22 or to another variant called model-based reasoning (MBR).[♦] In such systems, and for all but the most unique products, designers at CADD workstations can extract components from a database of standard components (which may be supplied by the company, a vendor, or the industry), store them in an appropriate CADD format, and introduce them into a particular design synthesis. We can identify a unique *instance* of a product as a series of components that has been selected, adapted, and appropriately connected to other components in the current synthesis model. As before, the component descriptions in the CADD database contain – or have pointers

> Work on model-based reasoning (MBR) has centered on the use of function-behavior-structure representations. Unlike case-based reasoning, where the "case" is retrieved and used essentially "as is," models allow inferences and different kinds of reasoning such as abstraction, prediction, and evaluation. This MBR research does not normally include solid models to represent structure, although they do not preclude it: they often just rely on schematic descriptions of structure (Goel et al. 1997).

to – several kinds of descriptive information (e.g., geometry, material properties, manufacturing specifications, and contractual data) that can be used to support design, manufacturing, and operation of the product. Each component is typically represented in a single frame or object at the level of detail at which we wish to reason about the system. Furthermore, a single component frame can represent a subassembly consisting of hundreds of separate parts or just one part of a product. Objects in these product models incorporate geometric, physical, and administrative attributes of a product's components as well as their topological structure.

One of the advantages of this approach is that with the product models explicitly embedded in a KBES, we can deduce a substantial amount of knowledge about the roles that individual components (e.g., beams, joists, or columns) play in the functioning of a product by reasoning about the hierarchy for a particular product and the topological links among its components. Thus, by noting that a beam is *part-of* a structural subsystem and is *connected-to* a generator platform, for example, we might conclude that its role or function in the product is to provide structural support for the generator. This allows the beam in our system to deduce that it supports the generator; thus, it can reference the weight of the generator in computing its size.◆

> A *partonomy* is a hierarchy constructed of object descriptions connected by "part-of" relations. Unfortunately, for the representation of knowledge, "part-of" has a variety of meanings in English (e.g., a brow is part of a face versus a wheel is part of a car), and it is not transitive unless it is used carefully. Typical part–subpart relationships often strongly influence the decomposition of design problems (Liu and Brown 1994).

One of the principal advantages of the MBR and similar approaches to integrated KBES-based environments is that we can use generic component libraries represented as frames, whose attributes and behavior can be inherited by instances of the components in engineered products or systems. Such a library provides significant leverage because, among other advantages, it drastically reduces the number of new rules we would have to employ in a given application. As we design new products that incorporate generic components, we can use our already-captured knowledge of system behavior encoded in the inherited behavior of the components. The number of new rules that we will need to configure such new products will increase only linearly, or even more slowly if new products share knowledge with past products.

However, there is a potentially serious roadblock to the widespread use of the kind of systems we are talking about, based as they all are on these very rich object descriptions.◆ Virtually all of our experience with such systems has been with research or

> Other key issues that challenge knowledge-based systems, in addition to the amount of knowledge, are maintenance and consistency. For practical real-world, rule-based systems, there can be thousands of rules. The standard example, XCON (McDermott 1982), at one point had more than 10,000 rules and needed 30 people to maintain it. Before it could be added, each proposed new rule had to pass a battery of tests to show that it was consistent with existing rules and would add more capability. Frame-based (object-oriented) systems have similar problems.

The National Institute of Standards and Technology (NIST) Design Repository Project (Szykman and Sriram 2002; Szykman et al. 2000) was a significant effort to store product information in XML about artifacts, functions, forms, behaviors, and flows. Regli et al. (2010) describe a case study of archiving the semantics of engineering artifacts. The design repository effort started at the University of Missouri–Rolla (Bohm et al. 2005) depends on the functional basis (Wood and Greer 2001). It is now at Oregon State University and includes about 100 products and more than 6,000 artifacts. The repository has been in existence for more than 10 years and includes bills of materials, functional models, function-component matrices, design structure matrices, and photographs. It has supported research on analogical reasoning, risk assessment, biomimetic design, and concept generation.

The challenges of building integrated systems for capturing, representing, and using engineering design knowledge are due to volume (i.e., there is a huge amount), complexity (i.e., complex devices involving multiple domains), distribution (i.e., knowledge will be distributed), diversity (i.e., there are many different formats and possible types of content), compatibility (i.e., due to formats, ontologies, standards, and errors), context (i.e., rationale, history, assumptions, and cultural influences are not stored), and task (i.e., the type of design reasoning being done).

demonstration prototypes. When we think of describing all of the components of a large building – an office skyscraper, for example – it almost boggles the mind to think of all the classes, objects, and instances we would have to develop (in the jargon, *instantiate*). A number of objects measured in the (low, single-digit) thousands would rank among the most robust and largest systems that have been developed to date.♦ No one really knows how such systems would scale up in a real-world context where tens or hundreds of thousands of objects would be required to realistically describe a complete design. Thus, we need to give careful thought to the implementation issues that underlie large integrated systems.♦

5.7 Bibliographic Notes

Section 5.1: The "languages of design" are discussed by Dym (1991) and Ullman (1992a). Rule-based and object-oriented representation are explained in Dym and Levitt (1991a). Integrated computational environments for design are discussed in Dym, Garrett, and Rehak (1992) and Dym and Levitt (1991b). The KBES for configuring electrical service for residential plats is described in Demel et al. (1992). Logic programming and neural networks are discussed in many introductory texts (e.g., Winston 1993).

Section 5.2: This section is heavily based on the excellent discussion of the role of drawing in mechanical design that is given in Ullman, Wood, and Craig (1990). Classical design textbooks that stress graphical communication are Dixon (1966) and Woodson (1966); more recent texts that fail to mention graphics at all include Ertas and Jones (1993) and Walton (1991). The role of drawing and visualization in architecture is articulated in an excellent book on architectural reasoning (Stevens 1990). The roles that drawings play in the design process are described in Tang (1988) and Ullman and Dieterich (1987); their role in grouping information is described in Larkin and Simon (1987); and Herbert (1987) reported on the use of drawings in architectural design.

Section 5.3: Discussions of the actual historical development of feature-based respresentation are found in Cunningham and Dixon (1988); Dixon, Cunningham, and Simmons (1989); and Dixon, Libardi, and Nielsen (1989). Features as volumes removed in machining are discussed in Henderson (1984), Pratt (1984), and Woo (1983); in processing in Pratt and Wilson (1985); and as devices in Krause, Vosgerau, and Yaramanoglou (1987) and Murakami and Nakajima (1987). The application of features to different processes includes Vaghul et al. (1985) for injection molding; Libardi, Dixon, and Simmons (1986) for extrusion; and Luby, Dixon, and Simmons (1986) for casting. A review of more recent work on feature-based design in general and on feature extraction in particular is contained in Finger and Dixon (1989b). Work on and an architecture for feature-based design systems is reviewed in Cunningham and Dixon (1987), and current and future research issues relating to their development and application are detailed in Dixon, Libardi, and Nielsen (1989). Commercial versions of KBES and feature-based design systems include Concept Modeller™ of Wisdom Systems, Chagrin Falls, OH; Cost and Manufacturability Expert™ of Cognition, Inc., Billerica, MA; Design++™, Design Power, Inc., Palo Alto, CA; and ICAD System™ of ICAD Inc., Cambridge, MA.

Section 5.4: An overview of object-oriented programming and further pointers to the literature are in Dym and Levitt (1991a). A KBES front end for a CADD system for ladder design is described in Lukas and Pollock (1988) and summarized in Morjaria (1989). The KBESs DEEP and LSC Advisor were first described in Dym et al. (1992) and Dym et al. (1988), respectively. The structural interpreter is detailed in Balachandran and Gero (1988) and summarized in Coyne et al. (1990).

Section 5.5: The need for integrated descriptions was argued by Dym, Garrett, and Rehak (1992) and Dym and Levitt (1991b). A vertical integration of KBESs within the domain of architectural design was presented by Fenves et al. (1990), and a more horizontal integration for structural analysis and design was proposed by Jain et al. (1990) and Luth (1990). The LSC Advisor, an early example of a KBES that, sitting atop an architectural CADD system, integrated the representations from the two systems, is described in Dym et al. (1988). The CADD system in question was GraphNet™ of Graphic Horizons, Inc., Cambridge, MA. An architectural design system implemented in the logic-based language Prolog is described in Rosenman and Gero (1985). Our discussion of object-oriented descriptions to structural standards processing is adapted from the helpful discussion in Garrett and Hakim (1992). The structural, fire-safety, and building-design codes referred to in this section (and the next) are, respectively, AISC (1986), Lathrop (1985), and UBC (1988).

Section 5.6: The discussion of integration with product modeling is based on Hakim and Garrett (1992); a formal approach to product modeling is outlined in Eastman, Bond, and Chase (1991). Discussions of design intent and its capture and utility are found in Ganeshan (1993) and Ganeshan, Finger, and Garrett (1991). As part of a discussion of the need for integrated knowledge environments, a taxonomy of engineering knowledge is proposed in Dym and Levitt (1991b) and specialized to design knowledge in Dym, Garrett, and Rehak (1992). The discussion of concurrent engineering in design and manufacturing is adapted from Levitt, Jin, and Dym (1991). An example of a KBES-database integration is the KADBASE system of Howard and Rehak (1989). The first version of the PDES/STEP standard (for mechanical product models and for printed-wiring-board data) is given in NTIS (1989). The model-based reasoning (MBR) approach to integrated representation is fully described in Levitt (1990) and summarized in Dym and Levitt (1991b).

6 Representing Design Processes

We now complete our discussion of representing *design knowledge* by turning to the representation of the design process in terms of the languages of design that we identified in Chapter 5. We take a path parallel to that of the previous chapter as we show how AI-based problem-solving methods can be used to model the ways we solve design problems, as well as how we represent the kinds of design knowledge we use to solve design problems.

6.1 Classical Design Methods

In Chapter 3, we described several *prescriptive* models of the design process. Inherent in these prescriptions is a disposition toward *inductive reasoning* wherein we try to induce or infer a solution to a design problem by filling in incomplete information or knowledge. We might contrast this with a *deductive* approach to design wherein, in its most prevalent implementation, we deduce solutions to our current design problem from case studies of previous designs. As discussed in Section 6.2, there is a rough analogy here between using a set of rules to analyze the meaning of a set of symptoms or conditions, in which case we are reasoning deductively, and using a set of rules to work "backward" from a set of desired goals in order to find the conditions that would allow us to achieve those goals, given the rules under which we are operating. In the latter case, clearly, we are reasoning more inductively than deductively. The classical or traditional methods that we describe in this section are clearly informed by and support the inductive nature of the design-process prescriptions presented earlier (cf. Section 3.3).♦

Recent studies seem to show that the reasoning used by designers is not pure but rather opportunistic. For example, designers do not work top down only (from general to specific) or just use deduction. This resonates with the suggestion that design can be characterized as a mixture of the standard logical inference types – deduction, induction, and abduction – that vary as to whether they are truth preserving. Deduction uses a specific example to reason from a general statement to produce a result in a way that preserves truth: for example, *B* is a bolt, all bolts are strong, and therefore *B* is strong. Induction infers from many different examples

(continued)

113

(continued)

that a general statement about them might be true: for example, bolt1 is silver, bolt2 is silver, bolt3 is silver; therefore, all bolts are silver. Abduction works with a general statement and a specific example to produce a hypothesis that might be true: for example, all bolts are strong, B is strong, and therefore B is a bolt. So, although design reasoning might include all of the above, it has been thought that designing might be an exercise in abduction and that the resulting design decisions are hypotheses about satisfying requirements: for example, design A produces behavior B, B is the required behavior; therefore, A is likely to be the design.

We add a contextual note. The prescriptive descriptions and their underlying methods are, in one sense, traditional or classical because they have been around for some time. Furthermore, and especially in the context of the arguments made in this book, these methods are labeled as traditional because they predate the vocabulary and structure of design as it is developing through AI-based research into design. It is also interesting to note that the methods to be described here are found only rarely in American textbooks on design. They are, however, found rather consistently in books written in Europe, especially in England and Germany.

We begin our review of the inductive methods of the classical tradition with the construction of *objectives trees*. These trees represent articulations of our design objectives or goals, working down from the most abstract – or top-level – objective, which we put at the head of the tree. Or, equivalently, we can think of each subgoal as a node of the tree and the top-level goal as the *root node*. Figures 6.1 and 6.2 show two objectives trees developed by different teams of undergraduates in a first-year design course in response to the following project statement given them by a client:

Design a "building block" analog computer kit. Design a rugged, low-cost, easy-to-use analog computer. It should be easy to reconfigure so that it can model a wide variety of systems. The basic functional blocks that need to be implemented (e.g., addition/subtraction, integration, and so on) have, for the most part, been determined. The focus of this project, therefore, is on the physical layout of the system, the choice of materials, and ergonomic issues.

Note that the objectives or goals for a project are not the same as goals set up in the procedural plans or methods used to solve design problems. Although both project goals and plan goals can be represented in a hierarchy, and both become more specific farther down, problem-solving goals tend to describe parts of the problem to be tackled (e.g., select a length and a width for part X that are compatible with parts Y and Z), whereas objectives tend to relate to performance and requirements. That is, problem-solving goals are usually associated with the plans or methods used to achieve them.

We can see in these two objectives trees how the client's overall goals for the project were refined into increasingly detailed subgoals. In effect, the two design teams built two hierarchical structures, and their goals in doing so were to clarify what was wanted by decomposing the client's objectives into their component subgoals. Note how the objectives became more specific as the client's relatively abstract project statement was clarified and refined, much as we found when we worked on the design of the stepladder in Section 3.1, although we can also see that the two design teams stopped their objectives trees at different levels of specificity.♦

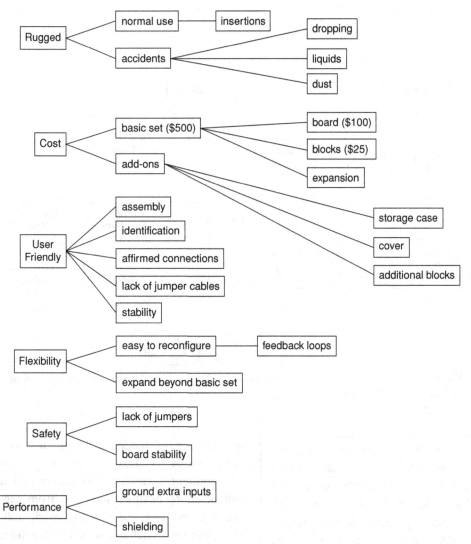

Figure 6.1. An objectives tree for a student-designed "building block" analog computer kit (Gronewold et al., 1993).

Another interesting feature of such objectives trees is that we can "read" in them the answers to two questions. As we work *down* the tree, we can answer the question of *how* to achieve the various objectives. Thus, the path outlining how the cost can be minimized is shown in Figure 6.1 as a decomposition of cost elements into basic attributes of the computer kit and "add-ons" – that is, extra features that add to the computer's price. These subgoals are then further decomposed as shown. Similarly, looking at Figure 6.2, if we want to know how to achieve the goal of portability, we see that we must achieve subgoals relating to the computer's weight, size, and power supply.

By way of contrast, moving *up* the objectives tree allows us to articulate *why* we wish to achieve certain subgoals. Thus, in Figure 6.1, we see that we wish

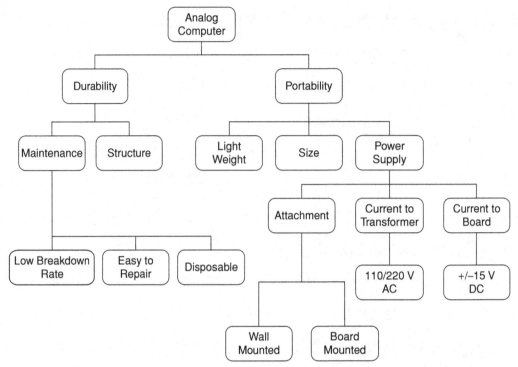

Figure 6.2. Part of the objectives tree for a student-designed "building block" analog computer kit (Hartmann et al., 1993).

Trees showing rule applications (for rule-based systems) also provide this simple form of explanation. Looking from a node representing the left-hand side of a rule toward the node or nodes representing the right-hand side can be used to explain *how* something was shown/decided. Looking in the other direction can be used to explain *why* (i.e., what goal/subproblem needed to be achieved or which hypothesis shown). Similar dependency structures between decisions also can be used during backtracking to help decide which decision to undo if a problem is found during designing (e.g., if a constraint fails). The assumption is that the current problem is due to a previously made decision. By undoing things in order (i.e., the last decision made gets undone first), we get (inefficient) *chronological* backtracking. *Nonchronological* backtracking allows us to undo a decision that caused a problem, wherever it is in the chain of prior decisions.

to incorporate feedback loops into our design so as to make the computer easy to reconfigure, which in turn means that it will be more flexible. Similarly, from Figure 6.2, we can see that we want reliable plug-ins to achieve a better structure, which in turn adds to the reliability of our design.♦

It is also interesting to comment on another aspect of choosing a direction in which to traverse objectives trees. We see in our discussion of AI-based problem solving in the next section that much of that approach can be cast in terms of searching a tree of possible problem states. Furthermore, the direction in which problem trees are searched has a meaning not unlike what we have just described for objectives trees. In searching problem trees, we often use a set of rules (cf. Section 5.1) to guide our search.

The direction of the search can be set in terms of matching data to the rules' left-hand sides, their antecedents, or in terms of matching goals to the rules' right-hand sides, their consequents, to see which data are needed to allow the goals in the consequents to be achieved. In the first case, we speak of data-driven, forward search. In the second case, we speak of goal-driven, backward search. Thus, analogous to objectives-tree traversal, we can think of (1) forward or downward-directed search as providing the data that explain how this path is unfolding, and (2) backward or upward-directed search as delineating the goals that explain why we are achieving these goals. We describe a problem-solving process called *means–ends analysis* in which the ends are the goals we wish to achieve and the means are the ways in which we actually arrive at the goal states. Means–ends analysis also has been proposed as a very general model of human problem solving.♦

> As we noted in a previous chapter, the AI technique of means-ends analysis is usually associated with recursively reducing differences between a current state and a desired state by applying actions.

Now, returning to the two trees shown in Figures 6.1 and 6.2, we see that they differ from each other in rather obvious ways, even though they were developed in response to the same client statement. However, this should not be a source of concern. The specifics of an objectives tree will vary with the inclinations and experiences of the individual or team that articulates it.♦ Clearly, the more specific we can be in developing the nodes of the tree, the more confident we can be that we understand our client's needs and can articulate an accurate design specification that expresses those needs. The point is not that objectives trees will generate a solution. Rather, it is that objectives trees are a useful way of translating and clarifying what our goals are, down to a level of detail that we need for the next steps in the design process. Thus, as we have noted before, objectives trees and the other aids we describe in this section are not algorithms that produce design solutions but rather aids that help us organize and externalize our thinking.

> There is a significant amount of research on the difference between expert and novice reasoning (Feltovich et al. 1997). Novices tend to use more textbook-like knowledge, for example, using standard hierarchies, and they search to find the right way to produce an answer. Experts tend to have incorporated or compiled more of their reasoning experience into their knowledge, so they more easily (with less searching) use the right knowledge at the right time.

Having identified our design objectives in greater detail by means of these trees, we are still left with much to do, including ranking these objectives according to perceived degree of importance. It is not unreasonable to argue that rankings ought to be deferred until design alternatives can be identified, in which case the ranking of objectives can perhaps be combined with a ranking of the utility or effectiveness of each alternative. However, it is equally plausible to argue that we should rank our objectives early on and use these rankings as a guide for focusing our design efforts. On this basis, we present now the ranking of objectives, which is implemented in a *pairwise-comparison chart*, as a tool to help rank the importance of the design objectives. As an illustration, Figure 6.3 is a comparison chart and its companion

Objective	Low Cost	Safety	Light Weight	Small Size	Reliability	Modularity	Total
Low Cost		0	1	1	1	0	3
Safety	1		1	1	1	1	5
Light Weight	0	0		0	0	0	0
Small Size	0	0	1		0	0	1
Reliability	0	0	1	1		0	2
Modularity	1	0	1	1	1		4

Figure 6.3. Ranking objectives (in a pairwise comparison chart) for a student-designed "building block" analog computer kit (Hartmann et al., 1993).

results for the analog computer design developed by the team that developed the objectives tree in Figure 6.1. We note that the objectives used in the comparison chart are somewhat different, both in terminology and in their respective depths in the tree, than are those in the tree. This is because the design team had several extensive discussions with the client between the time they prepared the objectives tree and the time they focused on the comparison chart. We can see that the chart is a relatively simple device in which we simply list the objectives as both the rows and columns in a matrix or chart and then compare them by pairs, proceeding in a row-by-row fashion. We assign a 0 or a 1 depending on how we assess the relative importance of each objective.

We can see from Figure 6.3 that safety is the most important consideration (viz., all the 1's in the second row) and that modularity is the second most important objective (viz., four 1's in the last row). Now we can list our objectives in order of decreasing importance as safety, modularity, low in cost, reliability, small in size, and light of weight.

Again returning to our objectives trees and proceeding downward into their depths, we also can use them to identify and characterize the functionality we expect of our design. That is, instead of simply viewing the trees as increasingly refined lists of attributes, we should begin to think of our design in terms of the functions that it is expected to serve, and we should extend this thinking to the designed artifact's components. A design aid called *functional analysis* may be helpful here. In functional analysis, we concentrate on *what* must be achieved by identifying and listing in an organized way the *inputs* to the designed device as well as its *outputs*. Our approach to functional analysis of a proposed device is to consider it first as a "black box" (see Figure 6.4(a)) whose inputs and outputs are defined at a fairly abstract level, consistent with clearly demarking the boundary between the device and its surroundings, much as in Simon's definition of design we recognized the designed artifact's inner environment as being distinct from the outer environment in which the device

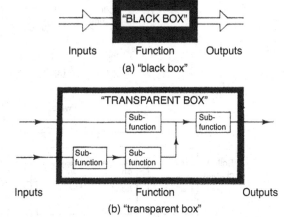

Figure 6.4. "Black box" and "transparent box" models of device functionality (Cross, 1989).

operates (cf. Chapter 2). In the next step, shown in Figure 6.4(b), we replace the "black box" with a "transparent box" in which we decompose the overall function into a block diagram of subfunctions whose composite functionality achieves our overall functional goal.

We can thus think of functional analysis as being equivalent to *function decomposition* – that is, to an attempt to parse the function of a complex object into subfunctions that can be achieved by individual components. Then, at this point, the design process becomes one of synthesizing an artifact by selecting and configuring components on the basis of their functionality, assuming that appropriate components can be identified. If not, they will have to be designed as well, and the functional-analysis process could proceed recursively.

We should keep in mind that one of the crucial aspects of doing a functional analysis or decomposition is a clear understanding of the system boundary and of all the inputs and outputs that cross that boundary. We illustrate this in Figure 6.5, wherein the inputs, outputs, and subfunctions are elaborated for a washing machine. Notice how the functionality requirements become more elaborate as we recognize the various inputs and outputs that cross the boundary over a washing cycle. Thus, at the gross level, we could say that the washing-machine output consists of clean clothes and dirt. In fact, however, at different stages, the washing machine will expel dirty water, less dirty water, and moist clean clothes. Furthermore, depending on our ultimate functional expectations, we could expect to produce both moist air and dry clothes if we shifted our boundary to include a drying function within our machine.♦

> For a review of recent functional reasoning research, see Erden et al. (2008) and the *AIEDAM: Artificial Intelligence for Engineering Design, Analysis and Manufacturing* special issues edited by Stone and Chakrabarti (2005). Umeda and Tomiyama (1997) provided a shorter, earlier overview.

Having identified the function(s) we wish our design to serve, how do we then choose the means by which these functions are to be effected? One useful tool for this stage of the process is the *morphological chart*, also known as the *function–means table*. In this table, we array all of the functions we wish to achieve against the particular means we can use to effect each function. We show an illustration for

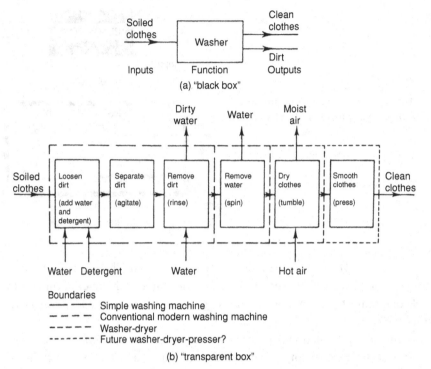

Figure 6.5. "Black box" and "transparent box" models of a washing machine (Cross, 1989).

the analog computer kit design in Figure 6.6. In the left-hand column are all the functions and subfunctions that the design must produce; in the row to the right of each function, we show or describe a number of means of effecting these functions. In principle, we can find possible solutions by adding or connecting a single means for each function listed, for example, by using an etched board for the block-to-block signal connection, relying on gravity to fasten blocks to the board, using concentric circles to connect power to each block, and so on.♦

The approach of moving from functional or behavioral descriptions to preliminary designs or embodiments (i.e., choosing component types that can provide the behavior and using them as building blocks) has been used in a variety of systems, including some of the functional reasoning systems referred to previously. Welch and Dixon (1994) used behavior graphs to model conceptual design: if the graph of a required behavior matches an existing embodiment (e.g., a gear pair), then that is identified as the solution; otherwise, the graph is transformed by adding intermediate behaviors. Larger graphs might match to more than one embodiment. This transformation and matching can be repeated until enough matches are found and a system configuration is formed. Although the control mechanism for that system is a fairly simple search, the A-Design system (Campbell et al. 1999) uses a knowledge-directed variant of evolutionary computing that is effectively a parallel search: a collection of candidate partial designs (a population) is altered by software agents that "grow" new alternative designs. Each agent "represents" a different electromechanical component type (e.g., gear) and tries to add it to partial designs in the population. The component types act as building blocks for the design. Once behavioral equations are attached to the resulting complete designs, instantiation agents can select actual components to replace the types in the configurations. The resulting actual designs are evaluated. Promising designs and promising portions of designs are kept and used to seed the next population.

We can clearly generate more design alternatives if we can identify a large number of means by which each function can be served. However, it is clearly a good idea to keep the list of required functions relatively short – perhaps in the range of six to eight functions or other attributes – so as to make the consideration of alternatives manageable in combinatorial terms.♦ It also is a good idea to express those functions at roughly the same level of detail or refinement so that both combinations and comparisons are of "apples and apples" rather than of "apples and oranges."

Function-based synthesis also can be done using a grammar (Kurtoglu et al. 2010), where grammar rules relate functions (e.g., from the functional basis), such as STORE electrical energy and SUPPLY electrical energy, to components (e.g., a battery) or configurations of components. This allows us to convert devices described in terms of "function structures" (i.e., configurations of functions) into configurations of components by applying rules. There are many research issues we must address about how to implement such rule-based systems, including acquiring grammar rules by inference from descriptions of known devices; using functional decomposition rules to convert an initial high-level, "black box" statement of the problem into a function structure; controlling how rules are applied to generate configurations; and developing and applying evaluation methods for the configurations. Cagan et al. (2005) survey grammar- and graph-based methods for computational synthesis, and Chakrabarti et al. (2011) also include other methods.

The design alternatives generated in a function–means table are not guaranteed to be admissible. The candidate designs may not make sense physically or economically, or they may not meet all of the constraints specified for our design. However, a morphological chart does provide a framework within which we can generate and explore alternative designs, which can in turn be tested for validity and utility. Thus, they support the idea of what in the AI literature is called *generate-and-test* problem solving (about which we will say more in Section 6.2). Similarly, morphological charts provide a systematic underpinning to the classic idea of solving problems by *trial and error* in that they provide an organized way of generating and listing trial solutions.

We have by no means covered all of the traditional or classical methods, only those we view as being most useful and having some relationship to the other process models we discuss later in the chapter. There are two other methods that are worth mentioning, however, if only in passing. Both methods depend on our ability to list, perhaps in a hierarchical way, the attributes and functions we desire of our designed device, and they are even less formal in their procedures than the methods just outlined. The first method is the *performance-specification method*, the idea of which is to develop a detailed list of desired attributes to which specific numerical values can be attached. These values then serve as target goals for the design process. The second method is called *value engineering*, and it aims to ensure that money is not wasted on unnecessary parts or components by asking whether each part serves a vital function and whether that part adds sufficient value to justify its cost in the design. This is done by listing all the artifact's parts and ana-lyzing the cost of each part as well as the value that part adds to the whole device.♦

It is difficult to perform intermediate evaluations of designs based on their cost (or other cumulative measures such as weight) because the cost of a half-completed design is usually not a good predictor of the cost of the final design. Experienced designers can probably conduct such evaluations in situations they know well. The MICON system (Birmingham et al. 1992) uses heuristics to address cost during designing, assigning penalty points to each design alternative based on the fraction of the allocated resource (e.g., available dollars) it uses, weighted by the fraction of the resource already used (i.e., a criticality measure). Although this strategy does not always work, and backtracking may be required, alternatives may be worse: for example, picking the cheapest choice at each turn may end up producing impossibly expensive choices later in the design due to interactions between choices. MICON is also a system that maps functions to configurations; however, because it focuses on high-level computer design, it includes both spatial and logical configuration (i.e., MICON decides how parts might be laid out on the board, not just how they are connected, by associating a "layout template" with each functional decomposition). So, when MICON decides that function X needs to be implemented with two subfunctions A and B, it also allocates space for them and identifies the required connections.

6.2 AI-Based Problem-Solving Methods

We now describe some problem-solving methods developed by AI researchers using programming techniques based on symbolic representation. With the advent of sym-bolic representation, as it became possible to think in terms other than numeric representation, we could manipulate concepts, objects, or "things," as well as col-lections of objects, such as lists of things. We present here a brief overview of these AI-based problem-solving methods, intending to provide just enough depth to make the balance of this chapter – concerned as it is with AI-based representations of the design process – coherent and intelligible. To illustrate the basic methodology and demonstrate the power and range of symbolic representation, we start with one of the classics of AI research, the *missionaries-and-cannibals* (M & C) problem, which we state as follows:

> Three missionaries and three cannibals seek to cross a river. A boat is available which will hold two people and which can be navigated by any combination of missionaries and cannibals involving one or two people. If the missionaries on either bank of the river, or en route in the river, are outnumbered at any time by cannibals, the cannibals will indulge in their anthropophagic tendencies and do away with the missionaries. Find a schedule of crossings that will permit all the missionaries and cannibals to cross the river safely.

In this rather wordy problem statement – perhaps not unlike some clients' design project statements – we have an *initial state*, six people on one river bank, and a *goal state*, all six having crossed safely to the other bank. Our immediate objective is to find a safe path between the initial and goal states. We introduce a set of *nodes* to represent allowable intermediate states of arrangements of the three missionaries

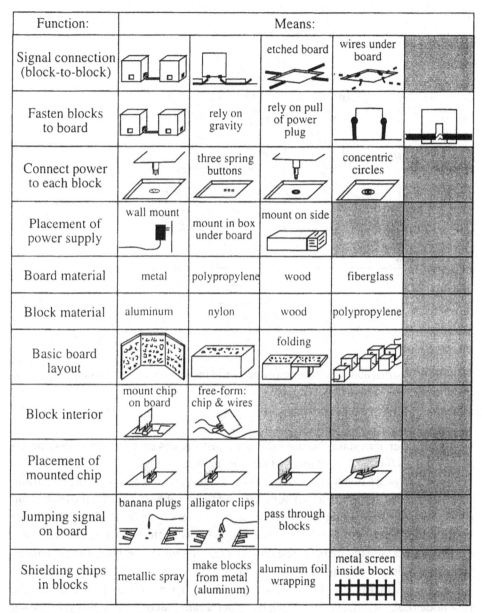

Function:	Means:				
Signal connection (block-to-block)			etched board	wires under board	
Fasten blocks to board		rely on gravity	rely on pull of power plug		
Connect power to each block		three spring buttons		concentric circles	
Placement of power supply	wall mount	mount in box under board	mount on side		
Board material	metal	polypropylene	wood	fiberglass	
Block material	aluminum	nylon	wood	polypropylene	
Basic board layout			folding		
Block interior	mount chip on board	free-form: chip & wires			
Placement of mounted chip					
Jumping signal on board	banana plugs	alligator clips	pass through blocks		
Shielding chips in blocks	metallic spray	make blocks from metal (aluminum)	aluminum foil wrapping	metal screen inside block	

Figure 6.6. Morphological chart for the "building block" analog computer (Hartmann et al., 1989).

and the three cannibals arrayed on the riverbanks, where the dashed line in each node represents the river (Figure 6.7). It would seem that 64 arrangements are possible because there are 6 people who can each be in one of two places. However, on reflection, we find that we need consider only what happens on one riverbank because each arrangement on that bank has a unique complement on the other bank. Thus, if we use the notation (#M, #C, +/−) to represent a state on the left bank – the sign indicates the presence or absence of the boat – it is clear that we can identify

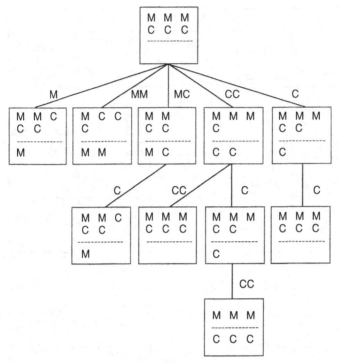

Figure 6.7. A partial node-and-link representation of the missionaries-and-cannibals problem (Dym and Levitt, 1991a).

only 32 independent states because there are just 4 states each for both missionaries and cannibals (i.e., 3, 2, 1, 0) and 2 states for the boat (+, −).

Now we ignore much of the detail of the solution we outline (e.g., the constraint that the number of cannibals can never exceed the number of missionaries) and concentrate on the representation. We represent the M & C problem as a set of states or nodes in the form of a *tree* and where transitions between the various states are represented by *links* (see Figure 6.7). The initial state is the *root node*. Nodes are *expanded* through the simple expedient of making all the moves that can be made in one boat trip, starting from the initial state of the root node. Five moves are possible from the root node (i.e., the boat can be taken by either a cannibal or a missionary, by two of one kind, or in one trip by one of each kind), and the resulting *child nodes* are shown in the tree's second layer. This process continues (cf. Figure 6.8) until an identifiable goal state – the state $(0, 0, −)$ – is reached, at which time the search can be concluded, the path back to the initial state can be retraced, and success can be announced.

The important point for us at the moment is that this amusing little problem is solved by searching a space of problem states, each of which is represented in a very simple – yet elegant – way. The representation helps us visualize both the states and permissible transitions, and it facilitates a very efficient statement of the arrangements and of the constraints. The representation also ignored a lot of irrelevant questions, such as "Is the width of the river important?" and "Are there

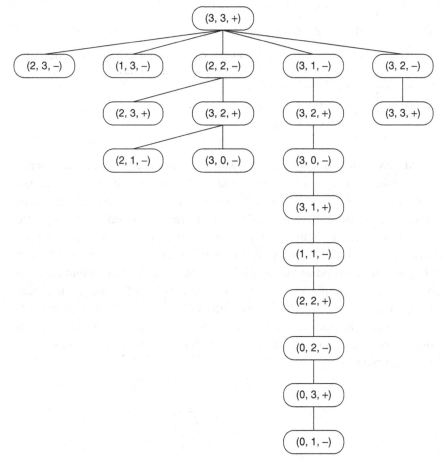

Figure 6.8. A partially developed search tree for the missionaries-and-cannibals problem (Dym and Levitt, 1991a).

restrictions on carry-on luggage?" Thus, we were able to set the representation at just the right level of abstraction for efficiently searching out a solution. The lesson is that we can, with the right representation, set up and solve many problems as searches through problem spaces.◆

Another classic search problem is based on a classic toy, the 8-puzzle (Figure 6.9), wherein we move tiles around in a 3×3 array or matrix containing eight numbered tiles and one empty space. It, too, can be set up as a search through problem states, as Figure 6.10 shows. We are interested here in one particular aspect of the complete solution. If we follow the paradigm introduced in the M & C problem, we can generate states and test

> It is appropriate to ask here whether design is search. In colloquial terms, designers move through alternatives gradually, making and refining choices until they find an acceptable design – and we can characterize this as search. However, we can model only the simplest design problems as a formal AI search that starts with a given: state space, well-defined goal nodes, start nodes, and operators. Designers produce new knowledge and requirements as they progress. Furthermore, design generally involves multiple spaces (see Section 6.2.1).

2	8	3
1	6	4
7		5

1	2	3
8		4
7	6	5

Figure 6.9. The 8-puzzle initial and goal states (Dym and Levitt, 1991a).

them to see if they match the goal state. However, a quick calculation of the combinations of possible moves reveals that there are 362,880 states that can be generated for the 8-puzzle – and this without accounting for copies of states or tracking moves so that they are not repeated uselessly! The search tree generated for the 8-puzzle is very wide; we can see in Figure 6.10 how quickly the tree broadens, and we are not even close to the solution. The 8-puzzle tree indicates perhaps more vividly than the M & C problem the need to understand the means by which we might traverse or search a tree. For example, do we expand the nodes by following the numbers shown above each node? Or do we follow paths such as 1–2–5–10 or 1–3–7–14? If so, why? Is there a difference between these two node expansion patterns? Given the breadth of this tree, can we invoke some knowledge to "narrow" it and so keep the search within bounds?

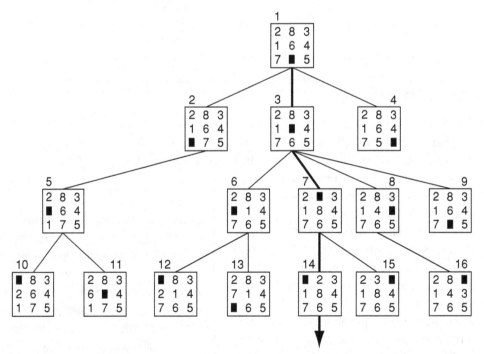

Figure 6.10. A partial solution to the 8-puzzle, where the search beomes very wide (Nilsson, 1980; Dym and Levitt, 1991a).

In fact, the 8-puzzle is an example of a search that can be *guided* or *directed* by applying some knowledge about the domain. We would *direct* a search by introducing an *evaluation function* that uses *heuristic knowledge* about the task at hand to assess whether a particular intermediate state is drawing us closer to our goal state. Thus, when we expand a node in the 8-puzzle, for each child node we would add the depth of the node in the tree to the number of tiles not in their proper places, and then follow the path with the lowest number.

What do our two sample problems have to do with modeling design processes? We want to build our vocabulary about trees as models of the search space, choosing a direction for searching a tree, problem characteristics, and problem-solving strategies for searching trees. Then we can describe some recent work in modeling the design process.

6.2.1 Models of the Search Space

We have already defined a *search space* or *problem space* as that space containing the set of states of the problem being considered, the *operators* that describe transitions between states, and the specifications of the initial and goal states. If each state or node in the space represents an actual configuration of the evolving solution, then the problem space is called a *state space*.♦ In the M & C problem, for example, nodes were first represented by the boxes in Figure 6.7 and then by the notation seen in Figure 6.8. The operators were shown as links between the nodes, and the links in turn stood for feasible moves that individual or paired missionaries and cannibals could make.

We can think of searching a state space as traversing a search tree, starting from the initial state or root node of the tree. As any given node is expanded, it is termed a *parent node*; its successors are called *children nodes* or *descendant nodes*. A node that has no successors or children is called a *tip node* or *leaf node*. Thus, the process of expanding a node is the process of generating its descendants. And, in keeping with our tree analogy, we talk about the *branches* of a search tree when we wish to refer to any nodes and links stemming from a particular node of the tree.♦

It is easy to believe that a "state space" contains all possible "states" in which a design (or configuration) may reside, including both incomplete and complete designs (Woodbury and Burrow 2006). However, designers often reason in multiple spaces. For example, when designing a mechanism, we traverse the normal design state space of complete and partial mechanism designs as we make further design decisions. We also might reason in the space of combinations of forces, rotations, and operating conditions that test the proposed mechanisms (i.e., a space of tests). Yet another space might be the suitable fixes we could apply if problems arise while we are designing. Still another space might contain design plans. The key concept is that we manipulate each space using different actions that act on different items and primitives.

Unfortunately, to break the analogy, we commonly talk about the "depth" of a search as we move along branches toward leaf nodes.

When we traverse a tree, movement among nodes occurs because some condition or set of conditions triggers an operator. Search operators are quite often written as rules, with the rule antecedents defining the conditions that must be met before the rule can be *fired*, and the rule consequents defining the actions to be taken if the conditions are met.

6.2.2 Choosing a Search Direction

When we use a tree representation, we have the option of choosing a direction for the search. We could search *forward* from our root node, *backward* from a goal state, or perhaps in both directions. Although many considerations enter into this decision (e.g., tree characteristics, problem and data structures, and problem-solving strategies), there are two basic strategies for making moves up or down a tree, and these strategies depend largely on the tree topology or shape. In one strategy, we match the given data against the antecedents of the available rules, find the applicable rule(s), and take the actions dictated by the consequents of the fired rule(s). This is *data-driven reasoning*, or *forward-chaining*.

In the second strategy, we examine the rule consequents to see which rules have as their outcomes the goals we wish to achieve. We then examine the antecedents of those rules that have the desired outcomes to see which facts we need for these rules to be fired. If the relevant facts are not available to trigger these rules, we take the establishment of those relevant facts as subgoals, and the search proceeds recursively in this fashion. This is *goal-driven reasoning*, or *backward-chaining*.

> Unlike basic search, more sophisticated and realistic search methods try to learn. So, the discussion of design processes ought to include the possibility of learning during or after designing (Duffy 1997). Grecu and Brown (1998) outlined several dimensions along which learning in design can be categorized, including what triggers learning (e.g., failure), elements that support learning (e.g., critiques), what is learned (e.g., design rules), the methods of learning (e.g., by analogy), and the consequences of learning (e.g., faster design processes). Brown (1991) presents a discussion of the use of learning to improve design process efficiency, and more recent work on learning in design is found in Duffy and Brazier (2004).

How do we decide on a direction for searching a tree?[♦] We normally look at two things: the shape of the tree as expressed in its *branching factor* (i.e., the average number of child nodes that can be reached from a given node) and the relative number of initial states. We would generally search in the direction of the lowest branching factor. If the branching factor is roughly the same in either direction, then our choice would hinge on how the number of initial states relates to the number of goal states, and we would work toward the larger number. Problems with a large number of goal states (e.g., chess, where the number of checkmates is quite large) are not amenable to backward-chaining.

However, whereas these principles are generally applicable to tree search, they may not be as useful for modeling a design process because other aspects of the process may come into play, such as the basic strategy we adopt for looking at one or another class of design tasks. It is also worth pointing out that backward-chaining is possible only if we have an explicit statement of the goal state (or of a rather small number of possible goal states). Furthermore, for problems where the goal is implicitly defined and, which is often the case in design, the solution is specified by various constraints, it is impossible to reason backward.◆

We also need to be somewhat careful with our nomenclature, or at least with the inferences we might draw from our terminology. In particular, although we talked before of working backward from a goal state, this should not be taken to mean that a goal cannot be defined as the root node in a tree, as we will see in our discussion of the PRIDE system in Section 6.3. In fact, still another characterization of problem solving is the contrast between *top-down problem solving*, which is usually meant to indicate a solution working downward from a high-level goal or an abstract statement, and *bottom-up problem solving*, which is usually taken to indicate an approach based on some original data or a very low-level problem description.◆

If we were given a design, then reasoning backward would be asking, given the requirements, whether we had the design knowledge to produce that design. But this is not the normal way things get designed! Backward reasoning makes more sense for diagnosis, where we are asking whether we could infer a hypothesized disease from the symptoms and any test results, using the available diagnostic knowledge.

Because humans are not limited to following a single control strategy and direction (e.g., top down) all of the time, computational design systems should also not be so limited. However, it is more difficult to include such variety in a system. At the very least, we need meta-reasoning to decide which strategy to use and when. Just as *meta-knowledge* is knowledge about knowledge (e.g., how reliable it is), *meta-reasoning* is reasoning about reasoning. In design, meta-reasoning might include reasoning about which plan to select or which analysis method to use (e.g., one is faster but another is more accurate). In rule-based systems, meta-rules are commonly used to decide which subset of rules to fire in a given situation.

6.2.3 Characterizing Problems

A few brief thoughts on characterizing problems are appropriate now, in anticipation of a subsequent outline of problem-solving strategies. Three broad problem types have been identified by AI researchers, of which two are of special interest here: path-finding problems and constraint-satisfaction problems.

The 8-puzzle and M & C problems are in the class of *path-finding problems* because the desired outcome is the determination of a set of moves, a path, from an initial state to a goal state. Engineering design problems can be set up as path-finding problems, as we see in the next section.

In many engineering design problems, we cannot state explicitly what form the solution must take, even though we can describe features or characteristics of the desired solution. That is, we desire in such problems to establish a goal state that can be described only implicitly at the start of the search. Such problems

Because design decisions must be both correct and compatible, almost all knowledge-based design systems include some type of constraint satisfaction, but rarely are such systems considered a standard *constraint satisfaction problem* (CSP) (Dechter 2003). Because variables can appear and disappear during configuration design, a slightly different approach known as a "conditional CSP" (at one time called "dynamic CSP") (Finkel and O'Sullivan 2011) is needed. When a new variable becomes active (e.g., because a part has been included in the configuration), the constraints that use that variable become relevant.

are called *constraint-satisfaction problems* because their solution depends on our ability to articulate, apply, and satisfy the constraints that form the implicit statement of our goals.♦

The problem characterizations we have just outlined are rather naive descriptions of problem solving. They are so abstract that they provide no guidance about how to find paths or satisfy constraints. Thus, we still need both strategies for achieving these objectives and algorithms (or rules or operators) to tell us how to expand the nodes in a search tree so that our searches can proceed. Furthermore, the strategies that we use to solve problems (see Section 6.2.4) may render as relatively moot the distinction just outlined, at least so far as choosing a problem-solving strategy is concerned. This is because the available problem-solving strategies often can be applied to both types of problems, and the differences between the applications become more of detail and philosophy than of significant principle. For example, although we lack explicit goal statements in constraint-satisfaction problems, as a practical matter, we have to generate solutions that we can test against these constraints to see if we are moving in the direction of our (implicitly stated) goal. When we are trying to construct a path between two states, we have to generate steps that we can take along the path and then test them to see if we are on the right path. Thus, in both instances, we need to generate candidate partial solutions, regardless of what kind of generic problem we are solving.

6.2.4 Strategies for Solving Problems

We now turn to a very important aspect of problem solving. Recall that even with our two relatively simple puzzle problems, we applied certain strategies even as we sought solutions although search. In the M & C problem, we introduced a strategy called generate and test, whereas in the 8-puzzle we talked about guided or heuristic search. Thus, problem-solving strategies are intimately tied to the ideas of search, and so it is useful to have a short discussion of some strategies for solving problems. The definitions and descriptions we provide are more detailed than the search-problem characteristics just described and can be used in either path-finding or constraint-satisfaction problems.

Generate and test may be the most basic strategy. As in the M & C problem, we simply generate all of the possible states in a systematic manner and then test each one to see if it is a goal state. A generator must be complete and nonredundant: it must generate every possible state – we cannot afford to miss potential solutions – but not more than once. We need a tester to evaluate each state generated to see if it is a solution or is at least moving us toward one. For problems having a large number of possible states, we should be cautious about using generate and test without any domain guidance because the search performance could be degraded by a combinatorial explosion.♦

We should also note a point to which we return in the next section – namely, that we can use generate and test as a strategy for generating solutions regardless of the direction of our search and independently of whether or not we have explicit statements of our goals. Thus, whether we are doing forward-chaining, goal-driven search, or constraint satisfaction, we can still use generate and test as a basic strategy within each of the contexts just mentioned.

Inasmuch as we apply no *domain knowledge* to aid the search process in its basic implementation, we consider generate and test to be *blind search*. However, as we hinted for the 8-puzzle problem, we can do a *directed* or *guided* search by using domain knowledge to *prune* branches of the tree that are unlikely to yield a solution. This technique also has been called *hierarchical generate and test*. We might also observe that function–means tables (or morphological charts) could be loosely viewed as hierarchical generate-and-test tools they provide a way of eliciting candidate designs within a framework heavily dependent on domain knowledge (cf. Section 6.1).♦

Generate and test (G&T) is appropriate if there is a very small search space (which is not likely in design), almost all points in the space are acceptable (they pass the goal test), there is no other good way to search, there is a good way to generate potential answers, and testing is efficient and effective. G&T can be improved by planning before generating (PG&T) (i.e., gathering information about likely solutions in order to guide the generation). G&T also can be influenced (and even improved) by allowing testing to indicate how good the generated hypothesis is and letting that influence subsequent generation. However, this runs the risk of getting stuck in an unproductive region of the search space. It also may be possible to move some of the knowledge inherent in testing back into generation so that generation becomes "smarter" (i.e., G&T becomes G$^+$&T$^-$). Of course, if all of the testing knowledge were put into generation, then no testing would be needed, and the system would only generate answers!

Hierarchical approaches search first at an abstract level and then descend to details. For example, hierarchical planning produces a complete abstract plan before refining that plan. There may be multiple "refinements" before we have complete details. Planning or designing *only* at the detailed level tends to proceed linearly with details at every step. There is usually no way to know whether these early detailed decisions are going to adversely affect later ones. This is especially true given how difficult it is to properly evaluate partial designs.

Another basic search strategy is *decomposition* or *problem reduction* in which we divide the problem into (presumably easier) subproblems. We are used to doing this with complicated integrals, for example, as we do when we make complex travel plans. We might decompose the mechanical design of a certain part into subgoals such as defining the geometry, checking stress and deflection values, and ascertaining its manufacturability. The process of subdividing goals is also

We have to be very careful when describing design decomposition to distinguish among decomposing goals, plans, and artifacts. Perhaps the most difficult thing about using the decomposition method is putting the solutions to the subproblems back together again and determining whether they are compatible: constraint failure is likely to occur if we have not addressed the subgoals carefully. It is usually a good idea to generate extra constraints or requirements and put them in place for each new subproblem so that our solutions are more likely to combine well.

Least commitment reasoning (LCR) means delaying decisions until we have enough supporting evidence to actually make them. Each new decision either adds new evidence or allows it to be inferred, which leads to opportunities to decide about some other part of the solution. For example, a choice of a material might add evidence about its required thickness or suitable fastening methods. We can easily add LCR to an opportunistic strategy in which control is given to whichever decision is ready to be made as soon as it is ready. Human designers tend to commit early based on their experience with previous similar design situations, which contradicts an age-old design adage: "Never marry your first design." It is true, however, that reusing a good previous design decision reduces work and probably helps make the new solution good too (as in case-based reasoning). However, early commitment prunes away other possible paths through the state space, removing the possibility of finding innovative solutions or even a globally optimal solution.

called *goal reduction* or, more casually, *divide and conquer*. Recall that we have seen an analog of this strategy in the classical approach to function decomposition (cf. Section 6.1).♦

If we can compare a current state to a goal state and describe the difference between the two states, we can apply a strategy called *match* in which the difference becomes the basis for choosing the next operator(s) to apply to the current state. Beacuse we need domain knowledge to assess such differences, we can see that match is a strategy of guided search. Match is also central to *means–ends analysis* because we need to compare differences between subgoal states if we are to assess where we are in relation to our goal state.

Two other strategies are also worth mentioning. The first is perhaps more of a statement of principle than a strategy whose implementation is especially obvious – that is, the strategy of *least commitment*. The idea here is exactly as expressed in the aphorism about marrying one's first design (cf. Section 3.2) – namely, to make as few commitments as possible to any particular configuration when the available data are either very abstract or very uncertain.♦ The other strategy is a deductive one – in contrast to almost all the others we have just described, which are essentially inductive – and is called *case-based reasoning*. The idea here is to see what we can learn or infer from prior cases and to see whether we can derive acceptable mutations from previous designs.♦

Case-based reasoning (CBR) is most appropriate when there are many cases available to cover the search space densely and uniformly, and where simple changes (adaptations) to a case are possible. If no existing case is close to the answer we need, or if the retrieved case needs major adaptations, then CBR's advantages are removed and a more complex kind of reasoning will be needed in to modify a case. The worst possibility is that modifying a case will be equivalent to designing from scratch!

6.2.5 Stopping Criteria; Satisficing and Optimizing

Our studies and work in the engineering sciences may have left us with the belief that engineering problems have unique solutions. This is because we often use linear mathematics – which imply uniqueness – to model physical problems. Furthermore, even for those problems having more than one solution, we are often content to accept any solution instead of searching for others – as we did in the M & C problem, for which we sought any safe path across the river. Whenever we are satisfied with any one of several solutions, without expressing a preference, we say that we are *satisficing*. We do this often in design because we have no mathematical algorithm or expression that expresses our design goals, so we are unable to find a formal optimum (see the following).

We can find solutions that satisfice simply by doing an exhaustive search until we find a solution. We could refine this process somewhat by collecting and ordering several solutions found by exhaustive search and choosing among them. We can take this idea to the limit by finding *all* the solutions and selecting the "best" from among them, using whatever criteria seem appropriate. Thus, we could have sought the *shortest* safe path (i.e., the one having the smallest number of moves between states) in the M & C problem. Here, we are obviously *optimizing*; that is, we are searching for the *optimum solution*.♦

Although optimization is a very important part of engineering design, we do not delve into it in this exposition because it is only indirectly connected to our present concerns. This is because the focus of research in optimization has been on achieving better algorithmic results rather than on integrating notions such as knowledge-guided search.♦

6.2.6 Weak versus Strong Methods; the Role of Domain Knowledge

We have several times referred to domain knowledge and its applications to search. Sometimes, for example, we do unguided, exhaustive search; that is, we simply expand every single node in a search tree without

> We can distinguish between the different kinds of results we can produce as we design. The crudest is the set of all possible designs, which includes solutions to other people's design problems as well as impossible or incorrect designs. Another, smaller set is of those designs that satisfy constraints: they will be acceptable designs but not necessarily for our problem. Designs that also satisfy the requirements will satisfice, to use Simon's term, because they are what we need. By imposing preferences (e.g., more steel in the design is better), we can get an ordering. By using some sort of equation to define quality (e.g., favoring low cost and low weight), we may be able to find an optimal design.

> Note that we can optimize designs with genetic/ evolutionary algorithms (Goldberg 1989; Kicinger et al. 2005; Mitchell 1998) in which a whole "population" of potential solutions is considered at once, using a fitness function to indicate how good each solution is. Different strategies (including computational versions of genetic crossover and mutation) are applied to produce a new population, with some bias applied to try to carry over fitter solutions into that population. New populations continue to be produced until no significant improvements in the best solutions are detected. Fitness is calculated in any way the implementer wants, from a simple equation combining cost and weight, for example, to using qualitative or quantitative simulation (e.g., how an artifact behaves during use).

invoking any special knowledge we might have about our design or problem domain. Searches done thusly are said to apply *weak methods* because they lack the power available through knowledge-guided search. Thus, their very generality implies a certain weakness. We also call these methods *syntactic search* because of the emphasis on form or grammar, lacking as they do the meaning or semantics offered by knowledge-guided search.

In fact, it might be worse than that. If, for example, domain knowledge is expressed in rules, each rule applies to a specific situation in a given problem. It will only work there, so it is powerful, but not general. Hence, we can apply a collection of rules only to a specific collection of situations. Furthermore, the use of the rules is binary: either they apply in that situation or they do not, without any graceful degradation. This is referred to as the "brittleness" problem (i.e., they work well within range but suddenly "break"). In contrast, human experts can fall back to general knowledge and additional weaker reasoning to overcome such brittle behavior.

In contradistinction to the weak methods, we speak of more powerful *strong methods* that incorporate varying degrees and kinds of heuristic and experiential knowledge. We are trading generality for power because although the weak methods are almost universally applicable, they may fail because of combinatorial issues. The strong methods, conversely, are limited in that the domain knowledge or heuristics used for a given problem are unlikely to be applicable in other domains.♦

When speaking of knowledge-guided search, we use the word *heuristic* to mean a rule of thumb or a piece of advice that we have learned from experience and that we know may not always work. Sometimes, as in the 8-puzzle and for means–ends analyses, we use a *heuristic evaluation function* to measure the (conceptual) distance between a given state and the goal state. Evaluation functions should provide good estimates of the merit of particular states in a computationally efficient manner. If an evaluation function is computationally complex, it may be more efficient to do an exhaustive search instead of spending computational resources on real-time evaluation.♦

A good heuristic for searches never overestimates the number of steps to the goal (Russell and Norvig 2010) and keeps the average number of branches generated at each step of the search as close to one as possible. Many situation-dependent heuristics are needed for design reasoning. Some heuristics might be general, such as making the most constrained choice first in order to prevent likely failure later, and others quite specific, such as selecting a preferred decomposition for the design of a particular component.

6.2.7 Procedures for Moving Through a Search Tree

Now that we have decided we can model our design problem as a search of discrete states, identified the design problem's characteristics, chosen a problem-solving strategy, decided what constitutes a solution, and reflected on how much heuristic or other domain knowledge we can bring to bear, *how do we actually move through a search tree?* What we are in fact asking is, "What procedures do we use to actually make a search happen?"

When we were looking at the 8-puzzle, we commented that we could follow the node numbers in Figure 6.10; that is, we could loop along on a row-by-row basis until we found a solution. Alternatively, we could choose one or more paths that went straight down and take them, in turn, down to their farthest point, stopping only if we happen across a solution as we descend downward. We have just described two classical, blind-search procedures: *breadth-first search* and *depth-first search*, respectively. In breadth-first search, we literally proceed by rows, and we hope that our solution will lie in a relatively shallow row. If not, we may expend substantial resources processing a tree – especially a broadening tree – both to expand every node in each row and to remember all of the previous rows' results when we start the next one.

In depth-first search, we dive down along some path, never knowing how deep it will go or whether there even a solution to be found along this route. Thus, the drawback to depth-first search is that we could plunge down into some very murky depths and not even reach a solution. In this case, we can backtrack up the tree to some convenient branch point and take the next alternate path. We could limit the damage that might be done by fruitless deep diving by setting a *cut-off depth* below which we will not go, even absent a solution. We should note that the memory requirements for depth-first search are far less extensive than for breadth-first search, although the processing time tends to be approximately the same in both cases. However, it can be said that when measured in terms of the path length between the root node and the solution, breadth-first search produces an optimal result – the shortest such path – whereas the same cannot be guaranteed for depth-first search.

Breadth-first search and depth-first search are the two basic types of weak search; that is, they apply no domain knowledge at all. They are also called *exhaustive* searches because they operate simply by following and exhausting lists of all the nodes in the tree as they are expanded, stopping only when a path runs into a dead end or a solution is found. At this level of description, these processes differ only in the way the lists are compiled. The weak methods can be improved by various techniques, but we find a real increase in power when we use guided or directed search. Here, we use heuristics, as in the heuristic evaluation function we described in the 8-puzzle problem, to speed up the search. For example, in a technique called *hill climbing*, we use a heuristic evaluation function to estimate the (conceptual) distance between a given node and the goal node, and we then expand the nodes in order of increasing distance; that is, we start with those having the shortest estimated paths to the goal. Hill climbing is a variation on depth-first search, and it is very much like local parameter optimization. It does have limitations; that is, it can attack local optima and miss the global optimum entirely.

There are other well-known guided search algorithms (see the bibliographic notes for further reading). Our point here is simply that they exist, and that they point to the exploitation of knowledge as a means for improving the problem-solving process. Although the kind of domain knowledge we might employ would appear to be

Most basic uninformed searches (i.e., using little or no domain knowledge), and even informed searches (i.e., heuristic), use a lot of memory (Russell and Norvig 2010), which makes them impractical for realistic, large-scale problems. For example, the well-known A* search minimizes the total estimated solution cost: it guides the search using the actual cost to a point in the search space, plus the estimated cost from there to the goal. However, an A* search records *all* currently explored branches, the number of which can grow exponentially as the search gets deeper. A "memory-bounded" version searches until the allocated memory is full, and then cannot add a new node to the search tree without dropping the oldest worst node. If every node in the tree can be used to get to n new nodes, and those n nodes each lead to n new nodes, then combinatorial explosion quickly overloads both space and time. Chess is a good example of the surprisingly large size of search spaces: it is estimated to have more than 10^{40} possible board positions. Smart control strategies and the use of heuristics are necessary to win this battle. One extreme example of this battle is a system intended to do creative design of electrical devices (Koza et al. 2004). It uses a form of genetic algorithm (called "genetic programming") to manipulate populations of alternative designs (e.g., for an amplifier). The authors report using a computer made of 1,000 Pentium II nodes running in parallel for 4 weeks in order to produce a solution – albeit an excellent and creative one that was patentable. In one run of a problem, the system considered more than 900 million individual designs!

different than algorithms such as hill climbing, conceptually we are attempting to do the same thing.♦

6.2.8 AI Problem-Solving Strategies in the Design Context

To complete this bare-bones introduction to AI-based problem solving, we now try to put the strategies we described into the context of design. For the present purposes, the context we use is a simple three-stage model of design akin to that described in Section 3.2 and in some of the taxonomies we set out in Chapter 4. Thus, we suggest the use of strategies for, respectively, conceptual, preliminary, and detailed design.

Conceptual design is, in a sense, the earliest stage of design, and at this point the design space is likely to be large and complex. This being the case, we prefer not to look for solutions in an ad hoc fashion because we may miss good solutions or, at best, take a very long time to find them. Thus, knowledge-guided search could be helpful. Decomposition or "divide and conquer," the reduction of a large problem into a set of smaller – and presumably easier – subproblems, is likely quite applicable. We have to bear in mind here that subproblems interact, so we must monitor individual solutions to ensure that we do not violate either assumptions or constraints of the complementary subproblems or of the overall problem. We should also try to limit our commitments to particular configurations because we do not usually have much reliable data this early in the design process. Knowledge from previous design cases can often be very useful either by suggesting solutions that have worked well in the past or rejecting those that have failed in the past. However, direct reasoning from past cases, called *case-based reasoning*, is more of a prospect for the future than the immediate present because there are many complexities in organizing and retrieving relevant cases for large, as-yet-unspecified spaces (even though we would love to learn deductively, as we do from case studies presented in the classroom).♦

When we get to preliminary design, we begin to worry about generating candidate solutions and testing and evaluating them. We test designs to ensure conformance with design objectives and constraints, and we evaluate designs against some metric (e.g., cost) that allows us to choose from among satisficing designs. Clearly, this is a time for generate and test, with the obligatory warning about combinatorial explosions, so we also must narrow the search space by pruning the search tree. We can achieve this end both by hierarchical pruning and by constraint propagation.

By the time we get to detailed design, we are fairly far down the design search tree. We understand whatever decompositions we have made, as well as their respective interactions. Consequently, we could characterize detailed design as being procedural in nature. Procedures for detailed design are rather rule intensive, especially in terms of the application of heuristics derived in large part from local experience. Much of the knowledge thus applied stems from an institutionalized understanding of past failures. This is oftentimes dangerous because we may be tempted to discard solutions that previously have failed, even if they failed for reasons that are no longer applicable.

6.3 Models of Design Processes

In this section, we discuss four knowledge-based systems – one in depth and three more cursorily – that emulate design tasks by applying some of the problem-solving strategies and methods just described. Of necessity, these descriptions are much shorter than the full descriptions available in the literature, so we may

---

Analogical reasoning has become increasingly important in recent design research (Goel 1997), especially for biologically inspired design (Chakrabarti and Shu 2010). Analogy is a special kind of knowledge-based reasoning that borrows from past solutions to inform the solution of the current problem. So, for example, the problem of storing cars in a small amount of space could be informed by the solutions to storing other things in a small area, for example, plates: stacks of plates can be mapped to stacks of cars, producing designs for high-density parking systems where cars are "stacked." Normally, some abstracted characteristic of the desired structure or behavior allows us to identify something from which we can borrow. For example, both electric current and water flow and roads direct the movement of cars, whereas canals direct the movement of boats. Similarly, pointers are long and thin and so are laser beams. Biology provides a rich resource for borrowing: armadillos, for example, might inspire a flexible armored coating for some device. Indeed, a whole new subfield of biomimetic design is emerging. We also can see from these examples why analogy is typically associated with creative conceptual design. For analogical design, it is possible to transform a previous design *solution* from one domain to another (e.g., plates to cars) or take the way that a design was *derived* (i.e., the design history) and modify that method to work in the new domain. We also can use analogical reasoning to simplify designs (Balazs and Brown 2002). We note that although analogical reasoning and case-based reasoning (CBR) are related, they are distinguished by the fact that analogies are usually made across domains, whereas CBR is not, as well as by the need for CBR to make only small adaptations to a retrieved case to fit the new problem.

In addition to the terms "conceptual," "preliminary," and "detailed," AI in design research has focused on "functional design," "configuration," and "configuration design" as well as on "parametric design." The first concentrates on producing a functional decomposition from the requirements (Erden et al. 2008). Configuration is concerned with arranging components selected from a given set of individuals or types, whereas configuration design has configuration at the core but allows components to be altered, usually by changing parameter values (Wielinga and Schreiber 1997). Parametric design focuses only on changing parameter values to meet requirements (Dixon et al. 1984). Although parametric design is usually detailed design, often concerned with dimensions, it is possible to do configuration design by carefully selecting parameters (Brown and Chandrasekaran 1989).

not do them the justice they undoubtedly deserve. However, our present purpose is simply to demonstrate that the approaches described in Section 6.2 can be used both to model real design processes and to further our understanding and vocabulary of design as a discipline.♦

The systems we discuss are as follows. First, in the domain of mechanical design, we describe a KBES that uses a generate-and-test paradigm to design V-belt drives. The second application is from the domain of structural design. The HI-RISE system does preliminary design of tall office buildings, using a generate-and-test strategy again, albeit for a more complicated problem. The third system we describe takes us back into the mechanical-design domain – in particular, the design of paper-handling systems in copiers. The PRIDE system extends generate and test to provide analysis and advice in the event of a design failure so that design failures can be rectified. Finally, from the manufacturing and assembly domain, we comment briefly on the R1 or XCON system that configures VAX computer orders for assembly on the factory floor. XCON, as it is now known, is perhaps the most visible success story of a practical application of KBES technology (and the representation techniques outlined in this exposition). We bring XCON into our discussion because its authors and developers intended it to use a generate-and-test strategy.

We perhaps should address the fact that all of our examples make us of the generate-and-test strategy. Is this all there is? In fact, they do not all share this characteristic because PRIDE adds to that strategy in rather significant ways, and we will see that R1/XCON in fact embodies a somewhat different strategy. However, it must be said that this is a prevalent strategy, both among designers for doing design and as a paradigm for modeling design in a computable form. It is partly that generate and test is very powerful, if the generation process is controlled properly, especially for routine and nearly routine design, where we have a lot of experiential knowledge that we can apply to help us generate sensible designs. It is also the case – as we observed in our discussion of AI-based problem solving – that we almost always wind up generating solutions and trying them out, even within the context of a more general view of problem solving, such as means–ends analysis. In computational terms, given that we can use rules to represent much of our experiential knowledge, and object-oriented representation to describe artifacts, their attributes, and their interaction, it would seem that generate and test is a natural implementation for routine and nearly routine design.♦

6.3.1 The Mechanical Design of V-Belt Drives

Our first description is of a system built to demonstrate an architecture, called the *design–evaluate–redesign* architecture, for AI-based systems intended to support certain styles of mechanical design. The principal idea behind this architecture is that design is an iterative process in which three tasks are often repeated – namely, evaluate a proposed design, decide whether it is acceptable and, if the design is unacceptable, redesign it. We note that different kinds of knowledge are required for each of these tasks. For example, the decision about whether a design passes muster is essentially a binary judgment – a design is either acceptable or not acceptable. However, for us to successfully redesign a failed device, we will likely require significantly more information than is embodied in the simple yes-or-no acceptability decision.♦

In the implementation of the V-belt design system, an algorithm was used to generate the original designs. For a given set of specifications (see the following discussion), an algorithm produced a complete design (cf. Figure 6.11). The testing and evaluation phases of the system applied classical utility-decision theory, whereas suggestions for the redesign were cast in terms of production rules. A frame-based representation was used to link these tasks (cf. Figure 6.12).

The original specification for a V-belt design problem is given in terms of values of six parameters. Of the six, four are hard or absolute constraints, for which very specific values must be achieved: drive speed, power, and the center distance limits. The other two

---

Note that in this chapter, "generate and test" (G&T) is often used in a more colloquial sense (i.e., things are produced and then tested to see if they are okay). In AI, G&T is usually associated with repetitively generating a complete hypothesis for a solution and then testing to see if it actually is one. Many computational design systems use "extend-model-and-revise" (EM&R) (Motta 1999), also known as "propose-and-backtrack." EM&R works by gradually extending the design one decision at a time, checking for problems at each step, attempting local changes (fixes) if they exist, and backtracking if no fix removes the problem. The key difference between G&T and EM&R is that the former produces and tests a complete design at each step, whereas the latter does not. "Complete-model-and-revise" (CM&R), also known as "propose-and-improve," starts by generating a complete (and preferably good) design and then trying to improve it, perhaps by hill climbing, without backtracking. Note that although DOMINIC uses CM&R, systems that use hierarchical heuristic design tend to use EM&R at different levels; for example, PRIDE and AIR-CYL (Brown and Chandrasekaran 1989).

The key ideas here are (1) "architecture," and (2) that different kinds of reasoning require different knowledge. The architecture is described by a task structure: a relationship between a *task* to be done, the applicable *methods* for doing it, the *types of knowledge* the methods require, and the *subtasks* they produce. Chandrasekaran (1990) suggests that a common task structure (the PVCM pattern) for design tasks is *propose* a solution (partial or complete); verify the proposal and exit if you are finished; *critique* the proposal to see what is wrong with it, and then exiting if it is okay; and *modify* the proposal to address the criticism. Because

(continued)

(*continued*)

each P, V, C, and M subtask can be implemented by a number of methods, combining them
enables a wide variety of possible ways to tackle
design problems. This is especially clear when we
note that some methods establish new tasks, which
themselves might have several possible methods.
Thus, PVCM also can be used recursively. Well-
established patterns are seen as "problem-solving
methods" (Brown 2009; Fensel 2000; Motta
1999) or "generic tasks" (Chandrasekaran and
Johnson 1993). One such task structure is encapsulated in the DSPL language that separates out
knowledge for selecting, decision making, failure
handling, and planning (Brown and Chandrasekaran
1989).

constraints, load speed and belt life, fall into
the category of having desirable goals as
their specification. In fact, designers strive
to achieve the maximum possible values for
both of these parameters. Figure 6.12 shows
two frames, the lower one being an instance
or *specialization* of the upper one, and we see
how some of the design parameters unfold,
as well as how the evaluation and redesign
are handled. Note in particular the redesign
recommendation in the last slot of the lower
frame.

The V-belt design system apparently
tested quite well in a comparison with
designs performed by experienced human
designers. In one cited example, it was given
the following design specifications: a drive
speed of 1,800 rpm, load speed of 1,200 rpm,
power of 40 hp, center distance restrictions of 12–28 inches, and required life of 4,000
hours. The algorithm produced a first design of a single belt (type 5 v), with load
and drive-pulley dimensions of 18.7 and 12.5 inches, respectively, a belt length of 80
inches, and a life estimated at 100 hours. The latter specification failure triggered a
series of redesigns, each having a life of at least 4,100 hours, the longest of which was
10,000 hrs. The final design consisted of four (type 5 v) belts, with load and drive-
pulley dimensions of 9.0 and 5.9 inches, respectively, a belt length of 50 inches, and
a life estimated at 5,360 hours. This design met all of the specifications and turned
out to be the cheapest as well.

We have indicated that the basic strategy outlined for the V-belt design system is
generate and test. In light of the suggestion we made earlier that generate and test is
almost always going to be a part of design problem solving, we could wonder whether
this basic strategy is so readily applied to design. In fact – as the authors of this system
also suggest – in its most basic form, generate and test will likely work only for

A design: Drive Pulley Diameter   D1
             Load Pulley Diameter   D2
             Belt Type   bt
             Belt Length   l
             Number of Belts   M

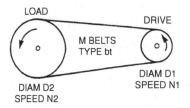

Figure 6.11. The V-belt design problem (Dixon, Simmons, and Cohen, 1984).

ATTRIBUTE	VALUE
Name	Generic V-Belt Drive
Specialization of	rotary-to-rotary power drives
Types	A, B, C, D, E
Input Specs	power (hp)
	load speed (rpm)
	center distance (in)
Evaluation Criteria	life (hours)
	max power (hp)
	installation clearance (in)
Name	V-Belt Drive
Specialization of	generic V-belt drive
Specification	power = 80 hp
	load speed = 1,200 rpm
	center distance = 15–19 in
Design	belts: 2 @ B75
	output: max power = 89.5 hp
	life: 7,500 hrs
	center distance: 14 in
Evaluation	power: acceptable
	life: acceptable
	center distance: unacceptable
Recommendations	increase belt length 4–6 in

Figure 6.12. Partial frames for the V-belt design system (Dixon and Simmons, 1983).

the design of components, parts, and small systems wherein the initial design and the iterated redesigns form the totality of the design picture. For more complex designs, where the decomposition of both the task and the artifact are more complicated, and more hierarchical, we need a more hierarchical approach to generate and test.♦

6.3.2 The Preliminary Design of High-Rise Office Buildings

The development of the KBES we now describe, HI-RISE, was motivated in part by a desire to externalize and record the

> DOMINIC is a classic example of complete-model-and-revise (CM&R) problem solving applied to non-decomposable parametric design problems where the values are numbers and hill climbing can be used to "improve" the design at each step. This system, although reasonably successful when compared to an expert, suffered from the standard problems associated with hill-climbing search (e.g., finding local rather than global maximums and becoming ineffective if it cannot find a hill to climb). As a consequence, a second version was produced with "meta-knowledge" that suggested ways to overcome those standard problems.

process of preliminary structural design – in particular, the process of establishing a structural configuration for tall office buildings. We are motivated to describe HI-RISE both because it is a design system and because it takes a more hierarchical approach to generate and test. Candidate designs for a structural configuration in HI-RISE are generated by choosing them from a set of feasible configurations, each of which is generated from a class of generic structural subsystems. The class of generic subsystems includes, for example, those that provide the resistance to lateral

Three-dimensional subsystems	core, tube
Two-dimensional subsystems	braced frame, shear wall, rigidly connected frame
Materials	steel, concrete
Linear components	beams, columns, diagonals

Figure 6.13. A static physical hierarchy in HI-RISE (Maher and Fenves, 1985).

loads on the building, such as loads due to earthquakes or wind. Subsystems that are used to carry lateral loads in high-rise buildings include rigidly connected frames, tubes, cores, and braced frames. In addition to a lateral-load resisting system, we also must specify a (horizontal) floor system to collect gravitational loads and a vertical system to carry those loads down to the building's foundation.

HI-RISE selects configurations and then tests them against several constraints, most of which are implemented as heuristics expressed as rules.[♦] Among the kinds of constraints that HI-RISE uses in its testing are spatial constraints on open areas, circulation, and the location of mechanical equipment; constraints on construction costs and time; functional constraints on the load path (via which loads are eventually carried to the foundation); equilibrium constraints (which are actually represented as Boolean functions); and strength stiffness constraints (also represented as Booleans but based on approximations of formulas from design codes).

The representation of the basic design information is hierarchical; that is, it is partitioned into a functional level and a physical level (Figure 6.13). HI-RISE starts by selecting a functional system, after which it proceeds depth-first to complete that functional system (i.e., lateral, horizontal, or gravitational) before attending to another functional system. The information at the physical level is itself organized hierarchically, being typically organized at descriptive levels such as three-dimensional, two-dimensional, materials, and linear components.[♦]

HI-RISE can be viewed as working with an AND-OR tree that gets more specific as you work down it. The OR nodes represent choices, whereas the AND nodes represent things that must occur together. Paths through the tree are selected using constraints to check that choices are compatible. HI-RISE can be seen as selecting and configuring compatible subsystems in order to produce a set of allowable designs.

Note that every system that has a hierarchical organization needs some sort of control strategy, probably heuristic, that orders the tasks at every level, deciding how to move through the hierarchy to schedule design decisions. This might be done by going depth-first to the most detailed decisions for every task in sequence, breadth-first by gradually moving the whole design to the next more specific level, or opportunistically by making decisions when and where they are possible (i.e., where the prerequisite conditions are met). The strategy can be controlled by selecting and using prestored plan fragments at every level, constructing plans during design, or setting necessary conditions as goals and detecting when they are met. Control choices (including plan selection) can be made with heuristics, means-ends analysis, or the weaker generate and test. A separate control strategy might be required for a rough design phase and for handling failure (e.g., constraint failure).

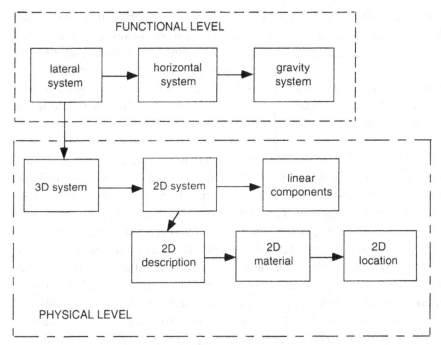

Figure 6.14. A dynamic physical hierarchy in HI-RISE showing four feasible design alternatives (Maher and Fenves, 1985).

The physical hierarchies are stored *statically* in HI-RISE's knowledge base (cf. Figure 6.13) and *dynamically* as a particular preliminary design is unfolding (cf. Figure 6.14). The dynamic physical hierarchy takes the shape of a tree. Each child node represents a feasible design developed to the next level of detail. We have completed a preliminary design for the specified functional level when we arrive at the bottom level of the tree. In Figure 6.14, we have four feasible designs for lateral-load systems: two have two-dimensional vertical subsystems and the other two have core structures: one concrete, the other steel.

The process of assembling or synthesizing a feasible configuration is, as we noted previously, a depth-first search through the static hierarchy for each functional system. Choices about contending alternatives are made by applying heuristic elimination rules; for example:

IF the number of stories is > 50
AND the three-dimensional system is a core
THEN alternative _____ is not feasible

IF the two-dimensional system
is a rigid frame
AND the material is concrete
AND the number of stories is > 20
THEN alternative _____ is not feasible

HI-RISE also does some analysis, albeit just enough to establish that a system can carry the anticipated loads, and it does select some parameter values in accordance with some heuristic approximations. For example, steel columns are usually taken to be W14 sections, whereas double-angle sections are usually specified for the design of braced diagonals.◆

> The configuration design is taken far enough to allow HI-RISE to evaluate the designs it finds and rank them for the user. HI-RISE is notable in that it provides more than one result, whereas most KBESs do not.

We can see in our abbreviated description that HI-RISE extends generate and test to a more complicated problem, the configuration of structural system for high-rise office towers. We also see here the extension to hierarchical generate and test, within the context of a data-driven or forward-chaining search. The data that drive the search are the original specifications for the building, which include a spatial grid that is entered manually by HI-RISE's users. The grid forms the set of spatial constraints to which the building must adhere; it includes topological constraints that define the number of stories and the number of bays in each plan direction, and geometrical constraints that define bay dimensions and minimum story height. Thus, like the system we describe next, the design begins with a spatial description, although we will now see how such a description initiates a goal-driven design process.

6.3.3 The Mechanical Design of Paper-Handling Systems for Copiers

We now describe the representation of a design process – for the design of paper-handling systems in paper copiers – within a knowledge-based system called PRIDE. This is a "real" system, now used daily as a design tool by designers doing feasibility studies for new copiers. It thus represents not just an academic exercise but also a convincing demonstration of the kinds of representation and reasoning that have been at the heart of our exposition. We do not describe the entire system here; we focus instead on some of the key points of modeling the design process.

If we were to look inside a copying machine, we would see paper being moved along a complicated paper path, past various components and physical processing elements, at fairly high speeds, and under rather stringent constraints. Whereas there are several kinds of paper-handling systems, the PRIDE system focuses on transport systems that use pinch rolls to move the paper. The design requirements are prodigious. They include geometrical properties such as paper entrance and exit locations and angles (Figure 6.15), timing requirements and constraints, permissible skew with respect to the path, tolerances on some of the engineering parameters, and the ability to adapt to a variety of paper properties, including size, weight, stiffness, and curl.

We can, in fact, decompose the design of a paper transport into subproblems, such as designing a smooth path between the input and output locations, deciding the number and location of pinch-roll stations to be placed along this path, designing a "baffle" to be placed around the paper path to guide the paper, designing the sizes

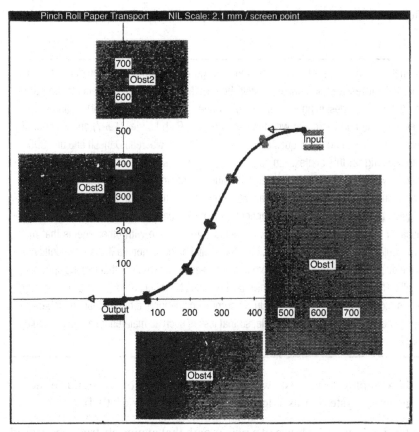

Figure 6.15. A snapshot of a sample paper path in PRIDE, including roll stations, input and output points, and obstructing regions to be avoided by the transport (Mittal, Dym, and Morjaria, 1986).

of various pinch rolls (drivers and idlers), selecting the proper materials for the pinch rolls and baffle, making decisions about paper travel speeds and the forces on the paper produced by the pinch rolls, calculating the time needed to move the various sizes of paper, and calculating the various performance parameters and ensuring that they satisfy the requirements. However, even with a very effective decomposition, the range of design tasks is enormously diverse because it involves making decisions about geometry, spatial layout, timing, forces, jam clearance, and so on – the totality of which are often beyond the scope of a single engineer. The same transport must be able to handle different sizes and weights of paper, which often presents conflicting constraints. For example, if the lengths (or widths) of the different sizes of paper are far apart, then the constraint on the maximum separation of neighboring roll stations for the smallest paper conflicts with the constraint on not having more than two stations guiding the paper for the longer papers. The design of the paper path is further complicated by obstructions that have to be avoided, as well as adherence to strict requirements on the smoothness, continuity, and manufacturability of the baffle in which the paper travels. All in all, paper-transport design is a complex domain.♦

Although complex to analyze and represent, paper transport design is quite restricted compared to most other design problems. Recent research is concerned with how to deal with larger problems, taking their cue from how people do it, such as designing with teams. Multi-agent design systems (MADS) can be used to model concurrent engineering and designing with teams in general. Both Lander (1997) and Shen et al. (2001) provide reviews of the issues and techniques associated with MADS, whereas Danesh and Jin (2001) address concurrent engineering. In this context, an "agent" is a piece of software that has its own goals, acts autonomously, and may communicate cooperatively with other agents to collaborate on the solution to a problem. Key questions include the following: How are agents controlled? Is the design description distributed or centralized? Can agents be developed to be independent so that they can be added or removed at will? How will work be allocated among the agents? How will agents communicate with other agents that may not share the same vocabulary or even the same ontology? How will agents negotiate if there is conflict? A "big-picture" issue is the decomposition of the design problem into pieces and whether that is done upfront or reasoned out dynamically. The pieces can vary from individual decisions to whole components. Another issue is to identify each agent's "point of view." For example, in concurrent engineering, different team members (or agents) can "represent" different phases of the life cycle, so that issues such as maintenance and packaging are all considered during design.

Given this complex design task, what does **PRIDE** do to be so helpful, either as a stand-alone design system or as a designer's assistant? In brief, PRIDE:

1. Facilitates the designer's choice of a planar path that avoids obstructions caused by equipment items within the copier and lies between specified input and exit points (see Figure 6.15).
2. Automatically checks that all constraints on the path geometry (e.g., smoothness and minimum radii of curvature) are satisfied.
3. Identifies a physical device (the baffle) in which the paper will be carried along the path.
4. Identifies, designs, and locates along the path the pinch-roll pairs that grasp and move the paper along the chosen path.

In fact, PRIDE simulates rather closely the actual design process that experienced copier-system designers have used for years. One of the results is that feasibility studies for preliminary copier designs are now completed and evaluated (with PRIDE) in hours rather than weeks.

PRIDE uses several representation schemes to incorporate heuristic, relational, and algorithmic aspects of the design problem, as well as several inference schemes at different levels of abstraction. PRIDE also has a powerful graphics interface that facilitates a rather complete simulation of the way human designers actually design paper-handling subsystems for copiers. Figure 6.16 shows a small part of an inheritance lattice that describes a paper-transport system designated as Trans5. In this object-oriented representation (cf. Section 5.1), Trans5 is an object with several attributes, some of which are linked to other objects, some of which are physical

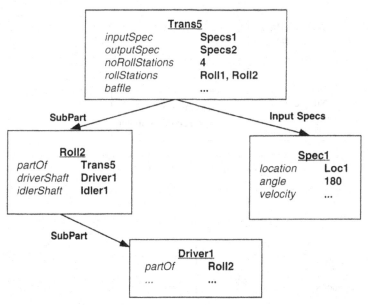

Figure 6.16. A stylized version of a small portion of the inheritance lattice that comprises PRIDE's knowledge base (Morjaria, 1989).

(e.g., Roll2, Driver1), and some of which are conceptual (e.g., Spec1). We see that Trans5 has components that are connected to it by one or more *SubPart* links (e.g., Driver1). When we wish to determine which specifications govern the input point, we obtain the answer through the *InputSpecs* link that defines the attributes of the design constraints at the point where the paper enters this subsystem (e.g., Specs1).

Already we can reason about the device we are designing (Trans5) because we can ask questions such as, "What are the output specifications that govern this design?" and "How many and what roll stations are there in Trans5?" We can see from Figure 6.16 that their answers are, respectively, Specs2 and 4: Roll1, Roll2.... With this representation for devices and parts thus established, we can focus on the process.

The knowledge base in PRIDE represents a design plan structured as a top-down process of identifying and satisfying design goals and subgoals (Figure 6.17). The design plan decomposes design goals into simpler steps. We see in Figure 6.17 a top-level goal of designing the paper transport, as well as goals for subproblems such as deciding the number of roll stations and deciding the diameter of the driver at Station 4. For our design plan to work, we must have the knowledge needed to order the steps, to perform each step, to detect failures in the design requirements, and to suggest fixes for the failures.♦

Note that PRIDE knows the steps to be taken in the task at each level of abstraction but determines the order dynamically. This stands in contrast to hierarchical systems that also use plans but where an order is specified. For most routine design tasks, experts do know the best order for subtasks (and the goals they satisfy) so that ordered plans can be acquired and used. In either case, there is no fixed overall plan for the design reasoning: high-level plans select and use lower-level plans, thus dynamically constructing a complete plan for the design at hand.

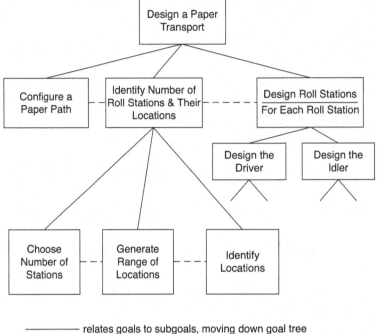

- relates goals to subgoals, moving down goal tree

- - - - - relates goals to their dependencies, moving to left from right

Figure 6.17. A stylized version of the goals and methods that comprise PRIDE's design plan (Mittal, Dym, and Morjaria, 1986; and Mittal and Araya, 1986).

We can think of the design process in PRIDE as one in which we are configuring a very complex system. We must first establish the geometry of a path along which the paper will be transported (see Figure 6.15). After the path is established, we (as the designers) must choose components and establish their configuration (e.g., size, location, and materials) so that different kinds of paper can be smoothly moved along the path, without jamming. The parts are chosen from a database of parts in normal use. Thus, in terms of the definitions and taxonomies discussed in Chapters 3 and 4, the process includes elements of both preliminary and detailed design. We can also characterize the process as routine design because we know how to decompose the design (cf. Figure 6.17), we know how to design the subsystems and components, and we know what to do when a constraint is violated.

Note that decomposition can be controlled with a plan; that is, with ordered or partially ordered lists of subtasks or subgoals, where each item in the plan represents a smaller piece of the problem. In situations where different plans are available, perhaps leading to alternative decompositions, we need to have knowledge available to evaluate the suitability of each plan to the situation so that we can select the best one. Decompositions also can be determined by reasoning, but this is more difficult because many different factors can affect how design decisions should be grouped together (Liu and Brown 1994).

We can think of PRIDE's problem-solving strategy being *generate–test–analyze–advise–modify*. It also makes very effective use of *decomposition* (as is quite evident in Figures 6.16 and 6.17)♦ and,

ATTRIBUTE	VALUE
Type	SimpleGoal
Name	Goal5
Descriptor	"Decide number and location of roll stations"
Status	INIT
AnteGoals	"Design Paper Path"
InputPara	"Paper Path," "length of PaperPath"
OutputPara	"Number of RollStations," "location of AllRolls"
DesignMethods	(SubGoals
	Goal51: "Decide min number of rollStns"
	Goal52: "Decide Abstract Placing"
	Goal53: "Generate Concrete Location"
	Goal54: "Build RollStn Structure")
Constraints	(Constr8: "First Stn <= 100 mm."
	Constr17: "Dist. between adj. stn <= 160 mm."
	Constr24: "Dist. between adj. stn >= 50 mm.")

Figure 6.18. The design goal "Decide number and location of roll stations" in PRIDE in its object-oriented representation (Mittal, Dym, and Morjaria 1986).

although we do not elaborate the details here, *constraint satisfaction*. We now describe just a few elements of the methods by which designs are generated and how failures are analyzed and fixed.

A design goal in PRIDE is responsible for designing (and perhaps redesigning) a small set of design parameters that describe some part or aspect of the artifact being designed.[♦] Some of the design parameters in this domain are paper-path segments, paper-path length, number of roll stations, diameter, width, and material of each pinch roll, baffle gap, baffle material, and time taken by each size of paper during transport. Figure 6.18 shows a simplified representation of the goal "Decide number and location of roll stations." The variables Descriptor and Name are used to describe the goal to the human users. DesignMethods is an ordered list of all the alternate methods for achieving the goal. In this example, there is only one method for carrying out this goal; that is, we must achieve four subgoals. Constraints contains the verification knowledge about the acceptability of a design.

Where does the "generate" come into the picture? In fact, one of the values of the slot DesignMethods could be a design

Although there are distinctions between types of design (e.g., conceptual and parametric), there is also a distinction between the phases of design (e.g., rough design, design, re-design, and redesign). Each needs different reasoning and different knowledge. Rough design is seen as partial or approximate design, perhaps only for some of the attributes, to determine whether a complete design might succeed with that approach. Design is the main event: trying to satisfy the requirements. Re-design is designing again, throwing away previous unproductive design efforts but preferably reusing the results of as much of the previous design effort as possible and reasonable. Typically, re-design uses a different method than before, incorporating new knowledge or newly discovered requirements gained from the previous design attempt. Although re-design might just as well be called "design," it is worth making the distinction in order to emphasize the difference in the activities and the context. Redesign involves incrementally changing a design in response to failure (e.g., executing a fix).

ATTRIBUTE	VALUE
type	InstanceSetGenerator
name	SetGen1
descriptor	"Generate standard driver diameters"
assignTo	(DesignObject defRollPair driver diameter)
initValue	"Find a diameter of 10mm"
classes	DriverDiameter
soFar	NIL
status	INIT

Figure 6.19. A simplified description of a method for generating the driver diameter by looking through a database of standard diameters (Mittal, Dym, and Morjaria, 1986).

generator, and the approach to generating designs could itself vary. We should note that from the point of view of capturing a lot of design alternatives, the design generators are among the most powerful design methods. These methods are all capable of generating different values for the same (or a small set of related) design parameter(s). We could attach to these methods heuristic knowledge for making "good" guesses about initial values to be generated. The generators in PRIDE also specify the ranges of possible values and increments. We show a design generator for the goal "design driver diameter" in Figure 6.19: it generates diameters for the drivers (in a pinch-roll pair) from a known database of acceptable driver diameters. This particular type of generator belongs to the class InstanceSetGenerator because the database is composed of instances of different classes of objects. The generated objects are instances of DriverDiameter. We also can see that this method specifies that 10 mm is a good starting value for the diameter, probably because the experts have found this to be a good default choice. Finally, it specifies that this instance object becomes the value of the design parameter driver diameter.

If our current design runs into trouble, if some requirement is not satisfied, the PRIDE problem solver analyzes the current partial design and tries to come up with suggestions to overcome any violations. These modifications may be heuristics reflecting a designer's experience in fixing similar problems, or they may be based on a more general problem-solving approach that analyzes dependencies between different parts of a design to suggest modifications that go beyond knowledge directly represented in its knowledge base. Figure 6.20 shows how advice is provided in PRIDE. In this example, the design goal "Decide number of roll stations" calculates a number of roll stations that produces a violation of the constraint on the maximum separation between roll stations. Advice – in this case based on a built-in heuristic – is provided to say that the number of roll stations should be larger than the number calculated.♦

Advice provided at failure time is also known as a "fix" or a "suggestion." For example, a fix might suggest a new value for a parameter, a range of possible values, or a direction in which to change the value (e.g., increase). Fixes can be predefined and stored associated with anything that can fail (e.g., a constraint, a method intended to provide a value for a parameter, or a plan selector), ready to be retrieved and used when that knowledge fails. Alternatively, fixes can be determined by analyzing the situation at failure time: versions of both PRIDE and the DSPL language provide this capability.

(continued)

PRIDE has many features beyond those we discuss here, among which is the capacity to maintain multiple designs simultaneously and to switch between different partial designs so that designers can explore different options in parallel. A designer can also selectively undo a design or impose additional constraints. In fact, PRIDE's many features make it very useful as a designer's assistant because designers working with PRIDE often develop suitable designs faster than either the system or the designer would have done alone.♦

This snapshot we have taken of the design process in PRIDE illustrates how a complicated configuration task can be described and analyzed with the aid of symbolic representation and concomitant problem solving. Furthermore, the stylized symbolic descriptions presented here contrast sharply with the numerical representations used in procedural programs. This clearly opens the door to detailed and structured

---

(continued)

DSPL provides a way to incorporate knowledge (known as "failure handlers") about which failure situations are worth trying to fix. The VT system (Marcus et al. 1987) also has predefined fixes, ordering them by the likely "damage" they could do to the existing design. VT cleverly provides an analysis in advance of using the system that detects fixes that might be antagonistic (i.e., that might lead to loops of alternating fixes in opposing directions). VT is also unusual in that it is able to combine fixes in order to recover from failure.

It is worth noting the difference that research goals make to the design and development of computational design systems. Although PRIDE mirrors the configuration design process normally used in that domain, it adds a lot of features intended to improve actual use. Other systems might focus on building

(continued)

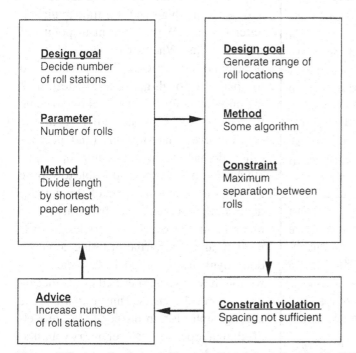

Figure 6.20. A stylized view of how design advice is handled in PRIDE (Morjaria, 1989; see also Mittal and Araya, 1986).

(continued)

the most realistic model of what a designer might actually be thinking and doing, producing the most optimal design possible, or using only one well-understood AI technique (e.g., a constraint satisfaction problem). Research systems tend to be built to help answer research questions about design knowledge or design reasoning. A wide variety of goals, and a wide variety of types of design problems and domains, makes it difficult to compare computational design systems.

discussions of design, both at the blackboard and on the workstation screen, because we can clearly realize both a vocabulary and a structure for talking about the design process.

6.3.4 The Configuring of Computers for Assembly

Our final discussion of a design system is also our shortest. We describe it only to make a point about keeping our mind open to different strategies for a given design task in much the same way that we advocate least commitment strategies in the context of adopting artifact designs. After all, a model of a design process is itself (or, at least, it ought to be!) the result of a conscious design process.

As we noted, R1/XCON is both a pioneering KBES and an extravagantly successful one.[♦] It is used on a daily basis in a very intense manufacturing environment. It is large, containing more than 10,000 rules, and it is maintained by a large staff (i.e., more than 50 people) because of the leverage that it provides its owner in configuring and delivering high-end scientific minicomputer systems. What we want to point out is simply this. When first conceived by its principal designer, R1/XCON was intended to operate through generate and test, with the initial data for a forward search to come from customers' computer orders. It turned out, however, that for this (and perhaps other) domain, *match* was a more appropriate strategy because it was possible to impose a partial ordering on some of the component configurations, so no backtracking or other failure recovery (e.g., redesign) was needed, and this partial ordering could be done dynamically – that is, "on the fly" – without having to be specified in advance by the system's designers. Thus, to paraphrase a design aphorism in the context of modeling design processes, "Never marry your first strategy."[♦]

XCON is not in use because the company that built it (Digital Equipment Corporation) is no longer in business. Before XCON was rewritten in 1987–88, about 30% of the rules were concerned with output or control; others were concerned with creating or "extending" partial configurations, database access, or computation. XCON and its other associated expert systems had about 10,000 rules and handled about 20 different CPU types and about 14,000 components – and all of these system dimensions were growing. It was 95% to 98% accurate, taking about 2 minutes per configuration, and saving the company some $4 million a year. It took about 30 people just to maintain XCON: about 50% of the knowledge in the system was frequently used, whereas about 40% of the rules changed every year.

XCON divided the configuration design process into sequentially ordered tasks, each completed before the next. The "match" strategy was used throughout, except for a "bin packing" portion of the problem where components had to be assigned to positions that provided resources (e.g., power): this could not be done without backtracking, so generate and

(continued)

6.4 Bibliographic Notes

Section 6.1: The classical design aids or tools outlined here are elaborated in the excellent exposition of Cross (1989), whose ordering and flavor we largely follow. Similar discussions are found in, for example, French (1985, 1992), Pahl and Beitz (1984), Ullman (1992a), and VDI (1987). The specific examples used to illustrate the methods (e.g., the objectives trees shown in Figures 6.1 and 6.2) were developed in a first-year design course taught at Harvey Mudd College (Dym 1993a; Gronewold et al., 1993; Hartmann et al., 1993). The idea that means–ends analysis is a general form of human problem solving was propounded in Newell and Simon (1963).

Section 6.2: The overview of AI-based problem-solving techniques is a condensation of the discussion of Dym and Levitt (1991a). The statement of the missionary-and-cannibals problem is taken from Amarel (1978), wherein the reader can find many elegant representations of that problem. The 8-puzzle is dealt with exhaustively in Nilsson (1980). A discussion of problem characterization, including the two-player games that are of no interest here, is found in Korf (1988). The meaning and utility of hierarchical generate and test are developed in Stefik (1990). Barr and Feigenbaum (1981) contains an extensive discussion of search, including definitions and algorithms for syntactic search and weak and strong methods. Search algorithms are also outlined in Charniak and McDermott (1985), Dym and Levitt (1991a), Stefik (1990), and Winston (1993). Problem-solving strategies are also discussed in Stefik et al. (1982). The notion of satisficing was first identified in Simon (1981). Optimization in design is discussed in Arora (1989), Fox (1971), Gero (1985), Paplambros and Wilde (1988), Vanderplaats (1984), and Wilde (1978). Case-based reasoning has only recently been applied in design; a good summary of its applicability to design is contained in Pu (1993). Gero (1987) addressed the issue of design mutations.

Section 6.3: The architecture for V-belt drive design is described in Dixon and Simmons (1983, 1984) and Dixon, Simmons, and Cohen (1984). The HI-RISE system was the subject of one of the earliest doctoral dissertations that applied KBES techniques to engineering design (Maher, 1984); HI-RISE is also described in Maher

---

(continued)

test was used. Within each task, the use of the "match" strategy ensured that each rule used kept the reasoning on the path to a correct solution (i.e., backtracking was not allowed). But to keep the reasoning on track, each rule required that many conditions be true before it could be used: about 30 on average. This made it very difficult to add a new rule because that required a correct, fully specified condition of applicability that was different from existing ones. This problem was exacerbated by the large number of people who were trying to maintain the knowledge base. A new methodology, known as RIME, was used to rewrite XCON to help with these problems. Each piece of knowledge was given a "role" (e.g., propose action, recognize failure), and different types of tasks were identified (e.g., select device, select location), allowing task-specific control. The new rules were less complex, related to fewer other rules, could be grouped, and allowed clear naming – all making it easier to add new rules. Although there were more rules overall, the new system's speed was comparable because it was easier to select rules and each rule took less time to process.

and Fenves (1985). Some of the earliest attempts to systematize approaches to design concepts and tasks is represented in the work of Brown and Chandrasekaran (1983). The most complete description of the PRIDE system appears in Mittal, Dym, and Morjaria (1986); further elaborations are given in Mittal and Araya (1986) and Morjaria (1989). The graphics interface to PRIDE is discussed in Morjaria, Mittal, and Dym (1985). The brief description of PRIDE presented here is adapted from Dym (1993b). The R1/XCON system is described in Barker and O'Connor (1989) and McDermott (1981, 1982).

7 Where Do We Go from Here?

We now bring our exposition to a close. We do so by first suggesting potential uses for and applications of symbolic representation in engineering design. We then offer a few remarks about research on representation in design, following which we conclude with some prescriptions for engineering design education.◆

> Many of the issues discussed in this chapter are still relevant. Because some of the newer aspects were presented in previous chapters, we try not to repeat them, but we will add a few more. It is clear from the changes to the topics addressed over time at the American Society of Mechanical Engineering (ASME) design conferences, particularly the Design Theory and Methodology (DTM) conference, that AI, computational design, and design cognition have made a *huge* impact. To see the extent of this influence and how far the field has come from being dominated by mathematical analysis (especially of kinematics), just look at some keywords and phrases, taken directly from abstracts of papers at a recent DTM meeting: cognitive model of design, design by analogy, Bayesian model, problem formulation in design, concept mapping, decision rationale, qualitative analysis of transcripts, integrated help documentation, design knowledge management, information foraging behavior, function modeling, nonfunctional requirements, inventive problem solving, object-oriented graph grammar system, interactive genetic algorithm, association rule learning, overcoming obstacles to creativity, conceptual sketches, model-based product development, identifying heuristics, addressing consumer variation, characterizing modularity, level of decomposition, design complexity metrics, reformulating ill-defined problems, designers' erroneous mental models, design fixation, biologically inspired design, emotion-based conflict, aesthetic preference, and consistency of human judgments.

7.1 Uses of Symbolic Representation in Engineering Design

We believe that the foregoing discussions confirm that there is considerable scope for applying symbolic (and other!) representation techniques in engineering design. In particular, we are thinking of the potential for using KBES techniques to represent both designed artifacts and design processes. There is a growing body of literature about and experience with both research prototypes and industrial applications that perform or assist in the performance of routine design; both R1/XCON and PRIDE are in daily use in their respective commercial domains. We reviewed some of the pioneering work in Section 6.3 – although we should recognize that in terms of engineering applications, this field is just over a decade old. However, we believe it is appropriate now to think about how symbolic representation and its extensions and applications could be used to model and assist with the diversity and complexity of the design tasks performed by engineers.

7.1.1 Integration with Analysis Tools

Although we have hardly mentioned the word in our exposition, *analysis* is an essential element of design. We use qualitative analysis in conceptual design to "get a feel" for things, approximate analysis in preliminary design to develop rough estimates and perhaps refine choices, and finally detailed analysis in detailed design to develop final values for part sizes, shapes, and so on.♦ Thus, one very effective application of KBES-based systems to design will be their use as interfaces or "front ends" to analysis packages. We can visualize them helping a designer choose analysis tools and interpret the results obtained from them. Analysis techniques (e.g., finite-element methods (FEM)) are often implemented as large programs that are not easy to apply, so KBES-based analysis tools could be used as expert checkers. An early step in this direction was the SACON rule-based system that provided advice on how to use the MARC™ FEM package. A later step was the MUMS research prototype, which explored some of the qualitative trade-offs we make in choosing among analysis strategies and tools at different stages of modeling in an engineering or design project. More recently, researchers have tried to clarify the role of and support analysis in the early stages of design, where modeling is a key issue.

We also can go beyond providing interfaces to analysis packages by integrating design and analysis knowledge in one knowledge base. We could then systematically check designs at each stage of the process instead of waiting for a stopping point (e.g.,

Qualitative reasoning can be used to get a feel for things before and during design. Symbolic qualitative values such as {high, medium, low}, {increasing, stable, decreasing}, and {solid, liquid} are used instead of numbers to identify the important distinctions in a domain. Then, differential equations, expressed qualitatively to use those symbolic values, represent how things may change. For a faucet, for example, equations link the angle through which the handle is turned with the size of the opening in the pipe, and link the size of the opening with the amount of water flowing. Such equations might indicate how the direction of one change *influences* the direction of another (e.g., increasing the opening size leads to increasing water flow), or state that the faucet does not change the water temperature. CADET, a case-based system for the conceptual design of electromechanical devices, is an example of a system that uses qualitative representations (Narasimhan et al. 1997). Qualitative equations are used together with causal relations (e.g., pressure difference causes flow) and influence graphs (i.e., nodes with directed arcs connecting them), which summarize the direction of the influences of variables on others. Influence graphs can be indexed by the type of variable (e.g., types might include flow, fluid flow, signal, or rotational signal), which enables fragments of larger graphs of existing devices to be reused in new devices. Bredeweg and Struss (2003) provide pointers to further literature in a special issue on qualitative reasoning.

the end). Furthermore, the reasons for the failure of the evolving design as determined by specific analyses can be used to provide suggestions for modifying and fixing the design (as in PRIDE), thus obviating the need to restart a design from scratch.

In fact, we really should be thinking about integrating in a seamless fashion all of our design tools, including all those used for analysis, graphics, documentation, accounting and cost analysis, part specification, manufacturing, and so on. However, this is still a daunting agenda, so we do not have much experience to report. One research prototype is IBDE, an integrated system for building design and construction, which represents a vertical integration of five task-specific KBESs:

1. ARCHPLAN, for assisting a user with the conceptual design of a building.
2. HI-RISE, for preliminary design of structural floor and wall systems.
3. SPEX, for preliminary and detailed structural-member design using formal representations of engineering design standards.
4. FOOTER, for preliminary layout and design of foundations.
5. CONSTRUCTION PLANEX, for developing a construction plan: identifying activities and their dependencies, estimating durations, estimating costs, and generating construction schedules.

We term this a *vertical integration* because each of the systems in IBDE conforms to the same underlying representation scheme, so the various KBESs have little difficulty exchanging data or results. By way of contrast, we can also imagine a *horizontal integration*, where the commitment to a description within a particular system – be it a KBES, a CADD system, or another component of the integrated system – is based on what is the best representation of the knowledge for that system (e.g., heuristic rules, a set of mathematical models, an algorithm, sketches or graphics, and so on). We would seek horizontal integration by facilitating translation among the representations so that other design knowledge can be shared as needed, in the representation that is most suitable for the immediate application of the knowledge. There has been some work along these lines in the domain of structural engineering, and some of the conceptual bases for such knowledge integration have been laid down, but there is not much else to point to at the moment.♦

> Computer-aided design and drafting systems can be augmented in many ways with qualitative or quantitative simulation (e.g., for mechanisms or for stress), manufacturability estimators (e.g., cost), planners (e.g., for disassembly or rapid prototyping), virtual manufacturing systems, checkers (e.g., for codes and standards), critics (e.g., to indicate potential life-cycle issues), constraints (e.g., for parametric and geometric design), and simple intelligence (e.g., smart cursors that detect what the user is trying to indicate or select).

7.1.2 Integration with Graphics Tools

In a similar vein, and as discussed in Section 5.2, we also should think about integrating our graphics tools with KBESs. To date, CAD systems have been largely restricted to documenting designs. Integrated CADD systems have been recently applied to ensuring spatial coordination among design components, although CADD systems acting alone have not as yet provided much support for automated design synthesis. However, the situation is changing, especially as more "intelligent" CADD systems are developed, as we mention herein.

We pointed out previously that sketches and drawings become, with increasing annotation, important sources of information about an unfolding design. Thus, tools in which we can integrate sketching and notetaking would be very valuable, especially if we can link them into CADD packages that can produce more formal drawings from our sketches. This also means that integrated KBES/CADD systems ought to support not only the graphics aspects of design representation but also feature-based approaches and others that make use of design knowledge in the terms that designers store and apply it. Thus, we would argue that another purpose for integrated KBES/CADD systems is to manage descriptions, functionality, constraints, and other aspects of design problem solving in a mode in which the representations are mixed and interchangeable, so that for the task at hand, the designer can use her system in the most natural representation as well as at the appropriate level of abstraction.

We have begun to see in the marketplace the emergence of commercial KBES/CADD packages (some are listed in the bibliographic notes for Section 5.3) that can be used to generate nearly routine product designs. These systems can be run in a fully automated manner or they can be used as design assistants by human designers with a variety of interface tools (e.g., geometric visualizations of the design). The (human) designer can then react to a given stage of the design and modify it as she or he sees fit. As we have noted, these tools are limited in their application to nearly routine, "semicustom" products for which a linear, sequential design process can be used.

The culture of design is changing; we more often see designers developing designs and syntheses on their own workstations. Because their designs are machine-readable even while still only partially developed, we have new opportunities to coordinate and control concurrent design in real time. An integrated KBES/CADD package could provide an alternative (to a red pencil on hard copy) for checking consistency, say, and communicating across disciplines electronically. Furthermore, models developed within a KBES/CADD package can be used by the package to check for spatial conflicts. This is why this form of automated verification of *spatial consistency* or interference checking has proven to be quite valuable to designers working in parallel. We can in some cases do away with building physical models of artifacts (e.g., scale models of refineries and full-scale mock-ups of new aircraft). However, because we still have trouble representing and reasoning about function (cf. Chapter 5), we cannot expect current KBES/CADD design tools to be very helpful in verifying *functional consistency* among an artifact's subsystems.

7.1.3 Collaborative and Concurrent Design

Section 5.6 discussed the importance of communicating about designed artifacts, thus perhaps referring for the first time to the communal nature of much design activity. In the "real world," complex systems often are designed by a team of designers and engineers (and others). This means that we must be able to decompose the complex system into smaller subsystems if we are to rely on a team design effort. It does not mean, however, that the subsystem designs can be carried out independently. A designer who is having trouble meeting the performance requirements for his or her subsystem often will "negotiate" with those who are designing interfacing modules in an attempt to modify or relax the requirements. Designers are constantly exploring choices and balancing different sets of constraints. Although we are still evolving models of computer support for communal interaction and concurrent design, we can safely say that extended KBESs can be used to model some aspects of communal activity. We present here a brief discussion of problems (cast in an organizational setting because that is what concurrent design is about) that could be opportunities for KBES applications, in the context of the elementary three-stage model of the design process (cf. Section 3.2).

In conceptual design, we expect that high-level specialists from all of the specialties involved will confer to identify the key performance specifications and constraints for each discipline and to ensure that assumptions used by designers from any one discipline – which may have implications for other disciplines – are communicated to and understood by all. Such coordination meetings, which can be viewed as tightly coupled concurrent design, constitute a serious overhead cost to the design budget, especially if the specialist designers work in separate organizations or are geographically dispersed.

Once we have decided on a specific design concept, as we start to embody particular schemes in preliminary design, we usually can parcel out the key subsystems to the appropriate design specialists. At the same time, we can identify the most important interface issues that must be coordinated with specialists in other disciplines. Once this occurs, design activities can proceed in parallel (or concurrently) in a fashion that we could characterize as loosely coupled. Few meetings are held, and the individual designers develop partial solutions covering the components, materials, dimensions, and other attributes of the subsystems assigned to them. We would see only limited communication across discipline- or subsystem-based boundaries.

When we get to detailed design, when final and detailed descriptions of all subsystems are determined for a large and complex artifact, we must share and coordinate a staggering amount of information. This is especially true for designed artifacts such as aircraft, power plants, and large software systems. Somehow, we have to guarantee that each of the artifact's subsystems will fit, both functionally and spatially. Traditionally, such coordination was achieved by circulating paper drawings to all affected parties, who would then "red pencil" conflicts, errors, or omissions. This process is clearly prone to error and very time-consuming because, as a practical matter, it is a linear process.

We can see that there are many issues involved in coordinating a design organization, depending on how interdependent are the specialists' tasks and on their novelty and uncertainty. Interdependence and uncertainty are always present in engineering. Thus, if we want to reduce the number of face-to-face coordination meetings, we might use KBES techniques to capture and apply the design-management heuristics of experienced designers and design managers. These heuristics help them to make some of the following kinds of decisions:

> *Defining efficient boundaries among the artifact's subsystems.* We want to minimize the number and severity of interface issues that must be coordinated across design specialties. Of course, defining such boundaries involves considerations other than ease of coordination (e.g., available fabrication and assembly methods, suppliers' capabilities, and so on).♦

An increasing number of researchers are studying modularity in product design. *Modularity* implies that modules are independent and that components within a module are similar. This enables us to develop product families in which modules can be interchanged on a common product "platform" to produce different members of the family (Simpson 2005). Current research on modularity focuses on heuristics, measures, and best practices (Zhang et al. 2004).

> *Making good guesses about the required degree of conservatism in the initial performance specifications for each subsystem.* We want to specify functional performance requirements and cost contingencies for subsystems so as to achieve a good balance between the costs of excess conservatism (e.g., wasted materials, labor, space, and power) and the potential costs of underdesign (e.g., loss of design flexibility, time lost to redesign, or increased probability of failure of the completed artifact).

> *Evaluating and limiting proposed changes in subsystem specifications.* We require experience-based knowledge to assess whether a proposed change is essential (i.e., the artifact will not meet an important specification without the change), desirable (i.e., the value of some performance enhancement will outweigh the costs of making the change), or undesirable (the costs of making a change will exceed its benefits).

> *Determining which specialists are likely to be impacted by changes that do get approved.* We ought to be selective in communicating information about such changes to the affected parties.

> *Controlling the timing and sequencing of that small number of key design decisions.* These key decisions, which often affect many disciplines, usually turn out to be highly constrained or constraining for one artifact or another.

Two architectures have been developed to model tasks carried out in parallel by human (and computer) agents. Both extend the style of KBESs in their complete dependence on symbolic representation and problem-solving techniques. These architectures are called, respectively, *blackboards* (BBs) and *cooperative distributive problem solving* (CDPS). These architectures, rather than rendering individual tasks into computable forms, focus instead on supporting the interactions of multiple agents because this is essential for the support of multidisciplinary design activities. We have neither the space nor sufficient background here to do justice to BB and CDPS architectures. Thus, we content ourselves with a few abstract characterizations of the two architectures and leave the interested reader to delve further into the literature.

The BB architecture attempts to coordinate problem solving much as do human engineering organizations. Thus, *domain knowledge sources* are discrete modules that incorporate chunks of individual expertise about some aspect of the design domain (e.g., organizational rules and procedures; design standards). The information-sharing aspect of face-to-face meetings is modeled by having the knowledge sources communicate indirectly with one another by writing to and reading from a common data structure called a *blackboard*. We should note here that the negotiation that happens in coordination meetings is not supported by this architecture. Finally, a BB architecture uses *control knowledge sources* to develop a strategy for solving a (design) problem. As problem solving proceeds, recommendations from the domain knowledge sources are evaluated against this strategy before being implemented. The control knowledge sources are intended to model the hierarchical nature of planning and supervision in an organization.

The CDPS architecture goes beyond the BB framework by allowing its equivalent of knowledge sources (KSs) to have considerably more autonomy than they do in a BB. In particular, in a CDPS network, we would replace each KS with a powerful KBES free to operate at all levels of the problem hierarchy. Furthermore, we would give each KBES the freedom to do its problem solving, based on its own knowledge and resources, without waiting to be ordered into action by a control KS. These ideas seem much more consistent with how we would do a complex engineering task in an organizational setting. We should note that BB architectures can accommodate CDPS ideas by using individual KBESs as control KSs that communicate with other KBESs through a shared BB space.

In CDPS, as we have hinted, problem solving is done by a network of loosely coupled, semiautonomous problem-solving agents or nodes. Each node in the network is capable of both sophisticated problem solving and cooperative interaction with other nodes. In fact, each agent may itself be a complex problem-solving system that can modify its behavior as circumstances change and plan its own communication and cooperation strategies with other agents. Thus, because each KBES in the network works on a piece of the problem, we are effectively decomposing the problem-solving process. We should recognize, therefore, that there are substantial issues involved in representing and decomposing the original global task without losing control of the solution process: we do want a coherent outcome! CDPS is a rapidly developing set of ideas that does seem suitable for managing multidisciplinary design. Further details are found in the literature cited in the bibliographic notes.

One final note on communal interaction and concurrent design: one problem common to many large design and engineering organizations is the use by

Agent-based systems also face the issue of inter-agent communication. Agent communication languages (ACLs), such as KQML and FIPA-ACL, have been proposed (Labrou et al. 1999). Messages have types (e.g., informing, querying, proposing, commanding, refusing) and can include "content" in any language that both agents share. KIF, the knowledge interchange format, is one such language: it allows arbitrary sentences in first-order logic to be expressed. ACLs support both negotiation and different architectures for multi-agent systems.

their staffs of many different kinds of databases, as well as other kinds of applications. This often is called the "Tower of Babel" problem because of the use of so many different languages and applications tools.♦ It clearly poses a major obstacle to computer-aided design integration. Of course, we could insist on the obvious administrative solution to this problem; that is, everyone on staff should use a single CADD package, although this assumes that the chosen package can support every kind of analysis needed. A KBES-style approach to the problem of multiple databases is based on the development of "intelligent" database management systems that can propagate and resolve constraints among related attributes of a design to specify additional attributes. Another line of attack is the development of knowledge-based database interfaces that can mediate between several incompatible but related databases.

7.1.4 Community Knowledge Bases

There is another dimension to the team effort – namely, how expertise is distributed among team members. For example, early in the development of the PRIDE project, while the expert designers' knowledge was being acquired, we found that each designer typically had some special knowledge about one or more design subtasks. It is easy to envisage that a KBES can be used to bring together within one system, "under one roof," this otherwise-distributed specialized knowledge. In effect, we create a community knowledge base that has more expertise than any single expert. One of the motivations for the PRIDE project was the establishment of just such a community knowledge base.

A related aspect is that a system that captures the knowledge of a design group or organization can be used both as a tutor for novice designers entering a design group and to propagate design styles and standards across an organization.♦ Novices can learn the design domain by exercising a designer's assistant on past or hypothetical cases, and they need to go to the human experts for help only when confronted with cases that are beyond the scope of the system. We think the propagation of design standards across an organization is also a self-evident benefit, and it is certainly feasible if the sponsoring organization is willing to invest in developing, maintaining, and distributing KBES-based design tools across their organizations. Perhaps the greatest benefit would be for companies whose activities are geographically dispersed because consistency of approach could probably be more easily enforced

It is clear that indexing, searching, matching, detecting relevance, and maintaining consistency across large distributed quantities of diverse data are significant challenges for the field of design whether it involves databases or the Web, both now and for the foreseeable future.

this way than with any other technique that would come to mind. Maintaining and periodically updating a quality design tool is likely to be much more efficient in this regard than is the periodic issuance of design directives in hard copy – copies that are likely to end up gathering dust in a file cabinet.

7.2 Research on Representation in Design

There is a very active community of researchers and designers who are exploring various aspects of representation in design. We will not review that work beyond what we have described in the preceding six chapters because, first, a detailed survey would likely be dated by the time our exposition appears. Second, comprehensive surveys are already available (see the bibliographic notes for pointers) and there are several meetings each year whose proceedings also provide current snapshots of what is being done. Thus, we limit ourselves to making a few observations about some aspects of KBES technology and about the style in which much of the research on design representation is being done. We begin our observations about style with some comments on KBES technology.

A major point of our discussion has been to link views of designed objects and the design process to their respective representations, a linkage that has been articulated and strengthened because of symbolic representation. We have made good use of this linkage, particularly as it has been reflected through the applications of KBESs to design modeling. Such KBESs, although they do increasingly involve object-oriented languages and concepts, are still reliant on rule-based technology. The reliance on rules as a form of expression provides several key advantages – many of which do not obtain for conventional, procedural programs – including having the ability to separate knowledge from the way it is controlled and used, making the knowledge used more transparent to the system user, making the solution process more transparent to the system user, and providing a capacity to add knowledge to the system incrementally.

Each rule (cf. Section 5.1) in a KBES can be viewed as a separate "chunk" of knowledge that specifies the situation in which it is applicable and the action that should then be taken. Because the consequences can be individually examined and understood, we say that the chunk of knowledge in such a rule is transparent. We might contrast this with procedural programs, in which we must look at the entire program to get a sense of the knowledge within because the knowledge used is inextricably intertwined with how it is used (i.e., its control).

Control in a KBES is exerted by its *inference engine*, which searches over a set of rules and keeps track of those whose situation clauses match the problem context at various points in the solution process. As a result, we can reconstruct how a solution evolves. This is why we talk about the transparency of a KBES solution. We can contrast this with procedural programs, which do not normally track their solution paths; they simply issue final answers.

Finally, because an inference engine can usually "figure out" which rules or knowledge to apply, we can develop a KBES *knowledge base*, in which the knowledge is stored, by starting with general rules and adding increasingly specific rules as we acquire more knowledge. We then expect that a KBES developed this way will exhibit expert problem-solving ability only after more specific, exceptional pieces of knowledge – expressed in terms of increasingly specific rules – have been added to its knowledge base. We do not see this capacity for incremental growth in most procedural programs because they support only one level of problem-solving knowledge.

We can summarize the three principal problems of rule-based systems as *narrowness*, that is, their range of applicability is limited; *uncertain coverage*, that is, they do not usually completely cover a problem domain; and *brittleness*, that is, they fail abruptly rather than gradually. The latter problem has been described as "falling off the knowledge plateau." Most AI researchers point to the need for richer, deeper representations of engineering knowledge for solving these problems.♦ However, we think that the issue is deeper than just representation.

To begin with, we know of no universally accepted definition of engineering knowledge that characterizes the individual components of this knowledge and indicates how they might be best represented. Knowledge is not a homogeneous quantity that can be uniformly represented by a single paradigm. Many different representation paradigms (e.g., frames, rules, procedures, formal logic, and graphics) are needed to adequately represent engineering problem-solving knowledge. However, we also need guidelines to relate representation tools to these different kinds of knowledge. Moreover, acquiring engineering knowledge for encapsulation in a KBES is a difficult, time-consuming, and error-prone process. It is likely the most important task in developing a KBES, as well as the one for which assistance – in terms of methods and automated tools – and deeper understanding is most needed.

> Rule-based systems are less popular. This is partly because they were oversold, along with KBESs in general, as the cure for all ills. Rule bases quickly get very large for real problems. They are also difficult to acquire and maintain. Rules may allow both consistent and inconsistent inferences; support the coding of the same knowledge in several ways; and incorporate a lot of situation-specific, not explicitly stated knowledge, which makes them very hard to reuse. Also, different rule bases in the same domain may not be combinable because they use different ontologies. However, rules can still be very convenient for small knowledge bases. Unfortunately, many of these issues apply to other types of knowledge representation as well. One approach to acquiring rules that removes some of the problems is to infer them from data: this is often referred to as *data mining*.

KBES developers have generally used one or, at most, a few knowledge representation paradigms to represent the knowledge for a particular problem. Knowledge that could not be represented was either "shoehorned" into an unnatural structure that worked only for that specific problem or ignored. Not enough attention has been paid to the organization of the knowledge. For example, it is likely that pieces of knowledge are missing, although not known to be missing, which may be why traditional rule-based systems are often brittle. We need, we would argue, a unified model of engineering knowledge, or else our system-building procedures are rather like those of "the blind men describing the elephant." A KBES developer describes the parts of the domain knowledge he needs to solve his narrow problem (i.e., the "elephant") without knowing how that knowledge relates to other components of the larger base of domain knowledge. Thus, attempts to integrate individual KBESs into unified systems then fail because each was developed in an ad hoc manner.

Thus, although we believe that the potential offered by symbolic representation has to a significant extent been realized in both our understanding of and our ability to do design, we also argue that we need further research into the nature of design knowledge (as well as into new representation paradigms). Thus, we echo the calls that have been made for a more systematic approach to design research, some of which inspired the taxonomies we described in Chapter 4. However, we do this with the explicit recognition that this cannot be done as an academic exercise alone, unconnected with real design activity. We wish to be quite clear that, as we said earlier in this exposition, our goal is not to automate design or to replace human creativity. However, if we better understand what design knowledge is and how it can be used, we can better develop design tools that help designers be more creative.

Perhaps another way of summarizing our argument is to note that for all the progress that is being made in representing – and, thus, helping us understand – the thought processes of design, we are in danger of loosing a flood of ad hoc systems of limited use. Furthermore, we are also in danger of developing a more arcane jargon, as opposed to a shared vocabulary and structure, because some of this work is being done without sufficient involvement of designers and other domain experts.♦ There has been a tendency by some AI-oriented researchers to develop models of how engineers and designers think by relying on superficial understandings of the underlying domains. We must remember that applying AI-based representations, as with any other techniques that are borrowed from other disciplines, is a *means* toward the *end* of understanding how engineers think, analyze, and design.

Now, as the work we described previously indicates (cf. Chapters 5 and 6), KBES technology is reaching a greater level of acceptance within the engineering community, the aforementioned drawbacks

> Messages in an agent communication language transfer knowledge, specifying the ontology that was used. This allows the receiving agent to make sense of the meaning of the terms used in the message, which facilitates any kind of knowledge use, such as sharing, reuse, and communication. People have tried the obvious approach of making "standard ontologies" for specific domains that can be adopted by many system builders: for example, Boeing and BAE Systems have tried to build engineering ontologies (Tudorache 2008). Such ontologies could be used by either human or computational agents. The most general approach is the "suggested upper merged ontology," which is intended to be a high-level ontology under which all other ontologies might fit (Pease 2011). For cases in which ontologies already exist, the ontology matching approach tries to make semantic correspondences between two different ontologies (i.e., In what ways are they similar?). This could then lead to a merging of ontologies. For example, an ontology about gears might be merged with one about belt drives to produce a third ontology or, alternatively, to produce a way of translating queries based on one ontology into queries based on the other.

notwithstanding. This acceptance is due to the fact that we can use KBES technologies to solve ill-structured problems by articulating and applying heuristic knowledge, whether that knowledge is experiential in origin or derived from first principles. But this acceptance could be limited if we do not recognize that we are dealing with a new paradigm of knowledge representation and modeling, and that there are formidable issues of verification and validation that must be addressed.

The *verification* aspect of KBES technology is concerned with whether the inference process is executing exactly as it is supposed to. Thus, the testing issues are largely domain independent and not too dissimilar from verification aspects of more conventional software engineering. The *validation* issue is concerned with the advice a KBES offers as it completes a task, whether acting as an automating adviser or as a designer's assistant. We ask a simple question: How do we know that the KBES has given valid advice? Now we move sharply away from conventional software engineering because our equally sharp departure from algorithmic approaches makes it almost impossible to set up benchmarks that will, in and of themselves, inspire confidence in our KBES. The difficulty is that we are dealing with a subjective assessment of advice that is produced by a nondeterministic, heuristic, and judgmental process. Perhaps the most convincing form of validation would be to exercise the system over as large a library of case studies as possible. Assuming the KBES provided reasonable answers, such an exercise might provide us with the confidence that comes from repeated successful experience. To be sure, there are questions about how many test cases we need to run and how accurate we expect the answers to be. However, we can likely invoke some statistics to assess the probability with which we can expect future results to be "valid."

The problem with the new paradigm is related to the issues of verification and validation, and it is simply that the old scientific test of repeatability does not easily apply in this domain. Because we are neither using the standard language of mathematics nor writing conventional algorithms that can be tested on standard benchmarks, we cannot count on other researchers and developers to repeat our analyses in their environments. Whereas in many ways AI research is almost experimental in nature, the kinds of programs that are written are not so easily specified or duplicated that they can be repeated as experiments in other researchers' laboratories. This may be one of the reasons that we never saw the AI boom that we expected in the late 1980s and early 1990s. Not only was the potential oversold, but we also may be victims of not having as yet a convincing enough paradigm for the way we do our research. The AI community is beginning to come to grips with this issue, but we must as well. Our own argument is that, as we have stated often, we can use this technology to develop a better vocabulary and structure for the discipline of design. We are not especially interested in automating design, save for those tasks which are largely drudgery. We are interested in providing support for designers, to free up their time and to enhance their creativity. There is some reason to believe that we are succeeding in this, and we will continue to do so as long as we maintain reasonable expectations.

7.3 Symbolic Representation in Engineering-Design Education

We describe in this book some of the ways that design can be discussed in a more coherent manner. We hope that we have demonstrated at least the beginnings of structure and vocabulary for the discipline of engineering design. We have tried to address a major concern of many engineering faculty members – namely, that design is "soft" and lacking in structure and rigor, by illustrating that there exist formalisms within which design issues can be systematically addressed. We argue that design is at least partially a cognitive activity that can be modeled. However, inasmuch as design knowledge is often design lore, the tools we need to develop a vocabulary and structure for the discipline of design must stretch beyond the languages of mathematics (e.g., numbers, symbols, and geometry) that have served us so well in analysis and design depiction. That is, we argue that the tools we need to develop the study of design should build on – but not be limited to – the traditional tools that are now the major focus of engineering education. We believe that some of these tools have come as a direct consequence of recent advances in the field of artificial intelligence, which have in turn offered us the opportunity to lay a better framework for design education and design research.♦

Papers in a recent American Society of Mechanical Engineering conference concerned with design education include these design representation and reasoning topics: studies of learning in and across courses, inspired by psychology (e.g., with pretests and posttests); building a case base of problems with their solutions; coding student activity using a function-behavior-structure ontology; the utility of using Wikis for shared representations; functional modeling; augmented reality; analogical reasoning for biomimetic design; sketch recognition; creativity enhancement techniques and studies; providing physical interpretations of mathematical expressions; and techniques for collaboration in teams. The case for increasing student awareness of representation issues has been made for some time (Dym 1999; Dym and Little 2009). Other related design education themes are found in Frey et al. (2010) and the references therein.

The tools of engineering analysis are well developed, powerful, and readily available in both analytical and computer expressions to those who wish to apply them. These tools are built on years of experience in the mathematical modeling of physical phenomena, often buttressed by experimental results or practical realizations. However, as any experienced engineer knows, there is much more to engineering – and especially engineering design – than can be captured in a formula or an algorithm. Until relatively recently, however, the tools for formalizing this "strategic" knowledge have not been evident. As engineering science has come to dominate the landscape of engineering education and engineering research, there has been no parallel development of vocabulary and tools for engineering design. Thus, as grounds for serious study, the "art of engineering" has lain fallow. To recognize that there is an art to engineering design does not preclude design from being worthy of serious scientific study. Furthermore, we now have tools to describe and delineate precisely those heuristic and judgmental aspects of design experience that we could not heretofore capture in the languages of engineering science. In short, an emerging paradigm based on symbolic representation and reasoning facilitates the systematic study of design in ways that the traditional language of mathematics cannot.

Thus, as we stated earlier, major motivation for this exposition is that the analytical modeling techniques that currently occupy most of the various engineering curricula do not represent a vocabulary complete enough for the synthesis task: that of generating and choosing among different designs. We are missing a language for representing design at a level of abstraction higher than that required for detailed design, for example, but with enough hierarchical structure to allow us to articulate at appropriate levels of detail all of the issues involved in making design choices. We believe that there must be a language or representation rich enough to span the gap between "design this structure to be stiff enough" and "place the rivets at the coordinates shown." The first statement is abstract and general, having limited practical meaning; the second is specific and detailed, coming when the design process is virtually complete.

We try to demonstrate in this exposition that recent developments in AI have provided new techniques for representing both designed artifacts and the design process, thus enabling better explication of design concepts. These in turn have enabled a more coherent structure and vocabulary for the discipline of engineering design. The kind of replication of a design process found in the systems illustrated herein cannot be achieved with traditional engineering-science or operations-research approaches, or in the graphics representations of modern CADD systems. These representations, although they permit the inclusion of economic, spatial, and other performance metrics, do not support qualitative or strategic issues that are not expressible in formulas or numbers. AI techniques thus offer opportunities to articulate design concepts, providing a better framework for design education and design research.

It is important for us to note that we are *not* suggesting that all students of engineering design become proficient in AI programming techniques. Neither are we adding one more requirement – and seemingly a severe one at that – to a curriculum already viewed as being both overburdened and lacking in depth. Our argument is simply that we are witnessing the formation of additional languages for engineering, which are further means for expressing the mental models that we use to describe engineering problems and their (physical) solutions. Choosing a representation is the beginning of modeling, although we have seen that representation in design is broader than in engineering science, where mathematical modeling is the key idea. Thus, our message is twofold. First, we should stress to a much greater extent to engineering students that mathematics is just one language for *modeling* physical phenomena and some aspects of the behavior of artifacts. That is, we must make students more conscious of the fact that engineering is about modeling and representing reality.♦

> This is an argument that was made previously (Dym 1999, 2004; McAdams and Dym 2004) and that others continue to make (McKenna and Carberry 2012).

Second, the approaches described in Chapters 5 and 6 offer *another language* for modeling various facets of engineering design. In fact, the symbolic representations exemplified there are not a second engineering language but perhaps a fifth. We are used to asking students to start their problem solving by describing the problem in words (the first language) to make sure they understand what is being asked of them. We suggest they then list all known quantities and their magnitudes, thus introducing numbers as a second language. We further suggest that students sketch the situation before writing down equations, thus introducing graphical representations before they get to the mathematical realization of the problem (the third and fourth languages, respectively). Thus, symbolic representation of artifacts and the design process could be viewed as a fifth language for engineering.

Now (and again), incorporating a fifth language does not mean that all engineering students learn AI. We have pointed to the emergence of commercial CADD systems that include some AI-based approaches in their design environments, so the power of symbolic reasoning can be made available to a designer in a user-friendly way. Students also can easily learn to think in terms of rule-based descriptions of the design process and in terms of elementary object-oriented descriptions. Inexpensive "expert system shells" are now available for many computers. These programming environments make it easy for students to play with and explore design – or other – engineering problems within the context of both design and analysis courses. Many of these tools allow students to use the *natural languages* of engineering in a natural way, blending technical terms and features with judgment calls and other heuristics. Such programming environments can be increasingly integrated with other computer tools for graphics, simulation, number crunching, and so on. Thus, we are getting closer to truly integrated computational environments for engineering modeling and computation, and we should let our students in on their development and their use. In so doing, we will strengthen engineering design as a discipline and strengthen engineering as a profession.

On the subject of teaching design, prompted in part by almost three decades of teaching engineering, we offer a few more comments.[*] It seems to us that we very often "lose the forest for the trees" when we implement ideas in engineering education. For example, how many students of drafting have been made to feel that the thicknesses of the lines on their ink-on-vellum drawings were more important than the nature and meaning of the drawings themselves? Too often, we focus on the

> Dym et al. (2005) present a broad survey of the many facets of not only design thinking and learning, especially representation and modeling, but also estimation, the design of experiments, the role of design teams, and the utility of intellectual and gender diversity.

details of execution rather than the ideas that we are trying to transmit. Now that we can use CADD programs to do all sorts of other fancy operations on drawings, isn't it time to introduce descriptive geometry? Now that we can take the drudgery out of drawing, shouldn't we focus on the meaning of drawings as sources of information?

Similar issues arise with the teaching of design. Currently, there appear to be three schools of thought in American engineering schools about teaching design. The traditional design school has it that design is experiential in nature, that "creativity cannot be taught," and that whatever discipline is imposed is done through scheduling and reporting requirements. This school also feels strongly that attempts to articulate and formalize a scientific theory of design will lead to the ruin of engineering-design education because creativity of necessity will be stifled. A second school of thought, unsurprisingly in view of the development of engineering education since the Second World War, is made up largely of engineering scientists and other "analytical types" who believe that there is no "real" content to design education. Reacting to the vagueness with which the content of design courses is discussed by the first school, the second school believes that no meaningful discipline of design can emerge until it can be put into mathematical terms. Just recently, a third school has emerged to propound the need for a scientific study of design as a cognitive activity that can be modeled within the framework of cognitive science.♦ Our own view is that the truth lies – as is so often the case – somewhere between these extreme poles. As we have said elsewhere, we

> Since the first edition of this book, there has been a dramatic increase in the number of empirical studies of how designers work (Subrahmanian et al. 2004). Many more researchers are analyzing protocols collected from designers and design teams during the design process (Gero et al. 2011) with the intention of revealing more details of design thinking. A better understanding of how design is done should enable us to provide far more effective computational support.

do not believe that we can model truly creative design – and this is perhaps where the debate has been miscast. Design activities encompass a spectrum from *routine* design of familiar parts and devices, through *variant* design that requires some modification in form and/or function, to truly *creative* design of new artifacts. Although we may not be able to teach creativity, we must recognize that the spectrum of design concerns does include many processes that are susceptible to thoughtful analysis – in other words, that are cognitive processes. . . .

. . . the root of the many complaints about the characterization of design as a cognitive process is due to confusion about where creativity and thoughtful process interact and overlap, on the one hand, and where they are distinct, on the other. This boundary is a moving one, especially in terms of our understanding. But we must be careful not to develop a new orthodoxy about design that prejudges where that boundary is and where, as a result, we preclude what we can learn and teach about design.♦

> The literature on stimulating creativity is vast: it seems reasonable to suppose that many of the techniques apply to designers generally (Shah et al. 2003a) and to creative design more specifically. Shah et al. (2003b) measured the effectiveness of ideation techniques using measures of the novelty, variety, quality, and quantity of ideas. Computational creativity has also attracted a lot of attention (Boden 1994), and computational design creativity is also now being investigated more seriously, although research has been underway for some time (Brown 2008; Gero and Maher 1993).

But even apart from this debate, we noted in Chapters 3 and 6 that there are several prescriptive models of design as a process and that they incorporate various kinds of inductive design aids and tools. As we also observed before, we can find few American textbooks on design that reflect even this level of thought about the design process (and we noted in Chapter 5 that sketching and drawing suffer from a similar fate). We do use the deductive method extensively because case studies are certainly a mainstay of design education. But for some reason, this collection of inductive tools, which students can easily learn and successfully apply (viz., the examples in Section 6.1), is largely found in European but not American textbooks. Our point here is simple. We believe that there is a discipline of design, that it encompasses much that can be taught to students to assist and channel their natural creativity, and that it is high time that we reform our engineering curricula to do so.♦

7.4 Bibliographic Notes

Section 7.1: HI-RISE and R1/XCON were discussed in Section 6.3.1 and references to them are found in the corresponding bibliographic notes. Some of the earliest attempts to systematize approaches to design concepts and tasks is represented in the work of Brown and Chandrasekaran (1983). PRIDE was discussed in Section 6.3

The continued health of the Design Computing and Cognition (DCC) series of conferences (previously known as "AI in Design"), as well as *AIEDAM: Artificial Intelligence for Engineering Design, Analysis and Manufacturing*, is a good sign that the AI, computational, and cognitive views of design will continue to affect design and design education. Recent topics in DCC conferences include design heuristics, the function-behavior-structure ontology, design agents, design rationale, enabling creativity, evolutionary algorithms, shape rules, brand identity, neural networks, belief models, automated layout design, clustering techniques, conceptual design of multidisciplinary systems, learning concepts, collective design, design teams, brainstorming, and biologically inspired design. There is strong interest in studying teams and finding ways to increase their effectiveness in modeling, studying, and enhancing creativity; functional representations and reasoning; analogical reasoning; sketch recognition; and support for biomimetic design. *AIEDAM* has recently had special issues on several of these topics, as well as on configuration, representing and reasoning in three dimensions, and the role that gesture plays in design. We should expect to see those topics, as well as other AI-inspired topics, appearing more in mainstream design and design education conferences in the not-too-distant future.

and the references to it are found in the corresponding bibliographic notes. Further engineering applications of KBESs are found in, for example, Brown and Chandrasekaran (1989), Coyne et al. (1990), Dym (1985), Dym and Levitt (1991a), and Rychener (1988).

The SACON system is described in Bennett et al. (1978) and Bennett and Englemore (1984). The MUMS system is described in Salata and Dym (1991). The application of KBESs to modeling in early design is discussed in Finn (1993) and Finn, Hurley, and Sagawa (1992). The IBDE environment is described in Fenves et al. (1988). IBDE's component systems are discussed in the following: ARCH-PLAN in Schmitt (1988); HI-RISE in Maher (1984); SPEX in Garrett and Fenves (1987); FOOTER in Maher and Longino (1987); and CONSTRUCTION PLANEX in Zozaya-Gorostiza, Hendrickson, and Rehak (1989). Elements of a horizontally integrated approach to structural engineering are found in Jain et al. (1990) and

Luth (1990). The arguments for knowledge integration are advanced in Dym and Levitt (1991b).

Our discussion of some of the aspects of communal and concurrent design activity is adapted in part from Levitt, Jin, and Dym (1991), which outlines architectures for concurrent design. The reference list of that article also provides an ample number of pointers to ongoing work in developing such architectures, but some good overviews are found in Nii (1986, 1989) for blackboard architectures and in Bond and Gasser (1988); Durfee, Lesser, and Corkill (1989); and Gasser and Huhns (1989) for CDPS. Engineering applications of distributed problem solving are described in Sriram, Logcher, and Fukuda (1989). The development of "intelligent" database management systems is described in Stonebraker and Rowe (1986). KADBASE, due to Howard and Rehak (1989), is an example of a knowledge-based database interface to multiple databases. The means that are used to coordinate engineering project teams are delineated in Logcher and Levitt (1979) and Thompson (1967).

The issues of using a KBES as a community knowledge base are explored in Mittal, Bobrow, and de Kleer (1984); Mittal and Dym (1985); Mittal, Dym, and Morjaria (1986); and Stefik (1986).

Section 7.2: An extensive and very thorough review of design research appeared in Finger and Dixon (1989a, 1989b). The volumes of Tong and Sriram (1992a, 1992b, 1992c) encompass a wide variety of current work, as do proceedings of relevant meetings, such as Gero (1992). Some of the concerns expressed about research on design representation were originally voiced in Dym, Garrett, and Rehak (1992). KBES architectures and features are described in Dym and Levitt (1991a). Lenat and Feigenbaum (1987) and Forbus (1988) discuss the limitations of rule-based representations.

Section 7.3: Much of this section is adapted from Dym (1993b). The quotation is from Dym (1992a), which was written following a spate of letters in ASME's *Mechanical Engineering* magazine in response to two articles by Dixon (1991a, 1991b) that proposed a more formal approach to engineering-design education.

References Listed in First Edition

A. M. Agogino, "Object-Oriented Data Structures for Designing by Features," prepared for the 1988 NSF Grantee Workshop on Design Theory and Methodology. Troy, NY: Rensselaer Polytechnic Institute, 1988a.

A. M. Agogino, Personal communication, 18 June 1988b.

AISC, *Load and Resistance Factor Design Specification for Structural Steel Buildings*, American Institute of Steel Construction, Chicago, IL, 1986.

C. Alexander, *Notes on the Synthesis of Form*, Harvard University Press, Cambridge, MA, 1964.

S. Amarel, "Basic Themes and Problems in Current AI Research," in V. B. Ceilsielske (Editor), *Proceedings of the Fourth Annual AIM Workshop*, Rutgers University Press, New Brunswick, NJ, 1978.

E. K. Antonsson, Personal communication, 25 January 1993.

L. B. Archer, "Systematic Method for Designers," in N. Cross (Editor), *Developments in Design Methodology*, John Wiley, Chichester, 1984.

J. S. Arora, *Introduction to Optimum Design*, McGraw-Hill, New York, 1989.

W. Asimow, *Introduction to Design*, Prentice-Hall, Englewood Cliffs, NJ, 1962.

M. Balachandran and J. S. Gero, "A Model for Knowledge-Based Graphical Interfaces," in J. S. Gero and R. Stanton (Editors), *Artificial Intelligence Developments and Applications*, North-Holland, Amsterdam, 147–163, 1988.

A. Balkany, W. P. Birmingham, and I. D. Tommelein, "A Knowledge-Level Analysis of Several Design Tools," in J. S. Gero (Editor), *Proceedings of AI in Design '91*, Butterworth Scientific Publishers, London, 1991.

A. Balkany, W. P. Birmingham, and I. D. Tommelein, "An Analysis of Several Configuration Design Systems," *Artificial Intelligence for Engineering Design, Analysis and Manufacturing*, 6 (3), 1992.

V. E. Barker and D. E. O'Connor, "Expert Systems for Configuration at Digital: XCON and Beyond," *Communications of the ACM*, 32 (3), 1989.

A. Barr and E. A. Feigenbaum (Editors), *The Handbook of Artificial Intelligence*, Vol. 1, William Kaufmann, Los Altos, CA, 1981.

J. Bennett, L. Cleary, R. Englemore, and R. Melosh, *SACON: A Knowledge-Based Consultant for Structural Analysis*, Report No. STAN-CS-78-699, Department of Computer Science, Stanford University, Stanford, CA, 1978.

J. S. Bennett and R. S. Englemore, "SACON: A Knowledge-Based Consultant for Structural Analysis," *Proceedings of AAAI*, 1984.

D. G. Bobrow and M. J. Stefik, "Expert Systems: Perils and Promise," *Communications of the ACM*, 29 (9), 1986.

D. G. Bobrow, S. Mittal, and M. J. Stefik, "Perspectives on Artificial Intelligence Programming," *Science*, 231, 28 February 1986.

A. H. Bond and L. Gasser (Editors), *Readings in Distributed Artificial Intelligence*, Morgan Kaufmann, San Mateo, CA, 1988.

D. C. Brown, "Routineness Revisited," in M. Waldron and K. Waldron (Editors), *Mechanical Design: Theory and Methodology*, Springer-Verlag, New York, 1991.

D. C. Brown, "Design," in S. C. Shapiro (Editor), *Encyclopedia of Artificial Intelligence*, 2nd edition, John Wiley & Sons, New York, 1992.

D. C. Brown and B. Chandrasekaran, "An Approach to Expert Systems for Mechanical Design," in *Trends and Applications '83*, IEEE Computer Society, NBS, Gaithersburg, MD, 1983.

D. C. Brown and B. Chandrasekaran, *Design Problem Solving*, Pitman, London, and Morgan Kaufmann, Los Altos, CA, 1989.

J. Cagan and A. M. Agogino, "Innovative Design of Mechanical Structures from First Principles," *Artificial Intelligence for Engineering Design, Analysis and Manufacturing*, *1* (3), 1987.

J. Carbonell, "Derivational Analogy: A Theory of Reconstructive Problem Solving and Expertise Acquisition," in R. Michalski, J. Carbonell, and T. Mitchell (Editors), *Machine Learning II: An Artificial Intelligence Approach*, Morgan Kaufmann, Los Altos, CA, 1986.

W-T. Chan and B. C. Paulson, Jr., "Exploratory Design Using Constraints," *Artificial Intelligence for Engineering Design, Analysis and Manufacturing*, *1* (1), 1987.

B. Chandrasekaran, "Toward a Taxonomy of Problem-Solving Types," *AI Magazine*, *4* (1), 1983.

B. Chandrasekaran, "Generic Tasks in Knowledge-Based Reasoning: High-Level Building Blocks for Expert System Design," *IEEE Expert*, *1* (3), 1986.

E. Charniak and D. McDermott, *Artificial Intelligence*, Addison-Wesley, Reading, MA, 1985.

W. J. Clancey, "Heuristic Classification," *Artificial Intelligence*, *27* (3), 1985.

Cognition, "Computers in Conceptual Design," *Computer-Aided Engineering*, May 1986.

R. D. Coyne, M. A. Rosenman, A. D. Radford, M. Balachandran, and J. S. Gero, *Knowledge-Based Design Systems*, Addison-Wesley, Reading, MA, 1990.

N. Cross, *Engineering Design Methods*, John Wiley, Chichester, 1989.

J. J. Cunningham and J. R. Dixon, "Designing with Features: The Origin of Features," in *Proceedings of the ASME Computers in Engineering Conference*, ASME, San Francisco, CA, 1988.

C. T. Demel, C. L. Dym, M. D. Summers, and C. S. Wong, *DEEP: A Knowledge-Based (Expert) System for Electrical Plat Design*, Southern California Edison–Harvey Mudd College Center of Excellence in Electrical Systems Technical Report, December 1992.

J. R. Dixon, *Design Engineering: Inventiveness, Analysis, and Decision Making*, McGraw-Hill, New York, 1966.

J. R. Dixon, "On Research Methodology Towards a Scientific Theory of Design," *Artificial Intelligence for Engineering Design, Analysis and Manufacturing*, *1* (3), 1987.

J. R. Dixon, "Feature-Based Design: Research on Languages and Representations for Components and Assemblies," prepared for the 1988 NSF Grantee Workshop on Design Theory and Methodology, Rensselaer Polytechnic Institute, Troy, NY, 1988.

J. R. Dixon, "Engineering Design Science: The State of Education," *Mechanical Engineering*, *113* (2), February 1991a.

J. R. Dixon, "Engineering Design Science: New Goals for Education," *Mechanical Engineering*, *113* (3), March 1991b.

J. R. Dixon, J. J. Cunningham, and M. K. Simmons, "Research in Designing with Features," in H. Yoshikawa and D. Gossard (Editors), *Intelligent CAD, I*, North-Holland, Amsterdam, 1989.

J. R. Dixon, M. R. Duffey, R. K. Irani, K. L. Meunier, and M. F. Orelup, "A Proposed Taxonomy of Mechanical Design Problems," in *Proceedings of the ASME Computers in Engineering Conference*, ASME, San Francisco, CA, 1988.

J. R. Dixon and C. L. Dym, "Artificial Intelligence and Geometric Reasoning in Manufacturing Technology," *Applied Mechanics Reviews*, *39* (9), September 1986.

J. R. Dixon, E. C. Libardi, Jr., and E. H. Nielsen, "Unresolved Research Issues in Development of Design-With-Features Systems," in M. J. Wozny, J. Turner, and K. Preiss (Editors), *Proceedings of the 1989 IFIP WG 5.2 Second Workshop on Geometric Modelling*, North-Holland, Amsterdam, 1989.

J. R. Dixon and M. K. Simmons, "Computers that Design: Expert Systems for Mechanical Engineers," *Computers in Mechanical Engineering, 2* (11), 1983.

J. R. Dixon and M. K. Simmons, "Expert Systems for Design: Standard V-Belt Drive Design as an Example of the Design-Evaluate-Redesign Architecture," in *Proceedings of the ASME Computers in Engineering Conference*, ASME, Las Vegas, NV, 1984.

J. R. Dixon, M. K. Simmons, and P. R. Cohen, "An Architecture for the Application of Artificial Intelligence to Design," *Proceedings of the ACM/IEEE 21st Annual Design Automation Conference*, IEEE, Albuquerque, NM, 1984.

J. M. Douglas, Personal communication, 1 August 1988.

H. L. Dreyfus and S. E. Dreyfus, *Mind over Machine*, Macmillan/The Free Press, New York, 1985.

E. H. Durfee, V. R. Lesser, and D. D. Corkill, "Trends in Cooperative Distributed Problem Solving," *IEEE Transactions on Knowledge and Data Engineering, KDE-1* (1), 1989.

M. Dyer, M. Flowers, and J. Hodges, "EDISON: An Engineering Design Invention System Operating Naively," in D. Sriram and R. Adey (Editors), *Applications of Artificial Intelligence to Engineering Problems*, Springer-Verlag, New York, 1986.

C. L. Dym, "Analysis and Modeling in Mechanics: An Informal View," *Computers and Structures, 16* (1–4), 1983.

C. L. Dym, *The Designer's Workbench: A Research and Development Proposal*, Internal Memorandum, Xerox Palo Alto Research Center, Palo Alto, CA, 1984.

C. L. Dym, "Expert Systems: New Approaches to Computer-Aided Engineering," *Engineering with Computers, 1* (1), 1985.

C. L. Dym (Editor), *Applications of Knowledge-Based Systems to Engineering Analysis and Design*, American Society of Mechanical Engineers, New York, 1985.

C. L. Dym, "Issues in the Design and Implementation of Expert Systems," *Artificial Intelligence for Engineering Design, Analysis and Manufacturing, 1* (1), 1987.

C. L. Dym, "The Languages of Engineering Design," *Engineering Clinic Orientation Luncheon Lecture*, Harvey Mudd College, Claremont, CA, September 1991.

C. L. Dym, Letter to the Editor, *Mechanical Engineering, 114* (8), August 1992a.

C. L. Dym, "Representation and Problem Solving: The Foundations of Engineering Design," *Planning and Design: Environment and Planning B, 19*, 1992b.

C. L. Dym, *E4 (Engineering Projects) Handbook*, Department of Engineering, Harvey Mudd College, Claremont, CA, Spring 1993a.

C. L. Dym, "The Role of Symbolic Representation in Engineering Education," *IEEE Transactions on Education, 35* (2), March 1993b.

C. L. Dym, J. H. Garrett, Jr., and D. R. Rehak, "Articulating and Integrating Design Knowledge," in *Workshop on Preliminary Stages of Engineering Analysis and Modeling*, Second International Conference on Artificial Intelligence in Design, Pittsburgh, PA, June 1992.

C. L. Dym, R. P. Henchey, E. A. Delis, and S. Gonick, "Representation and Control Issues in Automated Architectural Code Checking," *Computer-Aided Design, 20* (3), 1988.

C. L. Dym and E. S. Ivey, *Principles of Mathematical Modeling*, Academic Press, New York, 1980.

C. L. Dym and R. E. Levitt, *Knowledge-Based Systems in Engineering*, McGraw-Hill, New York, 1991a.

C. L. Dym and R. E. Levitt, "Toward an Integrated Environment for Engineering Modeling and Computation," *Engineering with Computers, 7* (4), Fall 1991b.

C. L. Dym and S. E. Salata, "Representation of Strategic Choices in Structural Modeling," in *Proceedings of the 1989 ASCE Structures Congress*, ASCE, San Francisco, CA, 1989.

C. M. Eastman, A. H. Bond, and S. C. Chase, "A Formal Approach for Product Model Information," *Research in Engineering Design, 2*, 1991.

K. A. Ericsson and H. A. Simon, *Protocol Analysis: Verbal Reports as Data*, MIT Press, Cambridge, MA, 1984.

A. Ertas and J. C. Jones, *The Engineering Design Process*, John Wiley, New York, 1993.

D. L. Evans (Coordinator, Special Issue), "Integrating Design Throughout the Curriculum," *Engineering Education*, July/August 1990.

S. J. Fenves, "Recent Developments in the Methodology for the Formulation and Organization of Design Specifications," *Engineering Structures, 1*, October 1979.

S. J. Fenves, "Expert Systems in Civil Engineering," Invited Lecture, *MIT Workshop on Microcomputers in Civil Engineering*, Massachusetts Institute of Technology, Cambridge, MA, 1982.

S. J. Fenves, Personal communication, 28 June 1988.

S. J. Fenves, Personal communication, 16 August 1993.

S. J. Fenves, U. Flemming, C. Hendrickson, M. L. Maher, and G. Schmitt, "Integrated Software Environment for Building Design and Construction," *Computer-Aided Design, 22* (1), 1990.

S. Finger and J. R. Dixon, "A Review of Research in Mechanical Engineering Design. Part I: Descriptive, Prescriptive and Computer-Based Models of Design Process," *Research in Engineering Design, 1*, 1989a.

S. Finger and J. R. Dixon, "A Review of Research in Mechanical Engineering Design. Part II: Representations, Analysis, and Design for the Life Cycle," *Research in Engineering Design, 1*, 1989b.

D. P. Finn (Guest Editor), "Early Analysis in Design," Special Issue, *Artificial Intelligence for Engineering Design, Analysis and Manufacturing, 7* (4), 1993.

D. P. Finn, N. J. Hurley, and N. Sagawa, "AI-DEQSOL: A Knowledge-Based Environment for Numerical Simulation of Engineering Problems Described by Partial Differential Equations," *Artificial Intelligence for Engineering Design, Analysis and Manufacturing, 6* (3), 1992.

P. Fitzhorn, "Toward a Formal Theory of Design," prepared for the 1988 NSF Grantee Workshop on Design Theory and Methodology, Rensselaer Polytechnic Institute, Troy, NY, 1988.

K. D. Forbus, "Intelligent Computer-Aided Engineering," *AI Magazine 9* (3), 1988.

R. L. Fox, *Optimization Methods for Engineering Design*, Addison-Wesley, Reading, MA, 1971.

F. Frayman and S. Mittal, "COSSACK: A Constraints-Based Expert System for Configuration Tasks," in *Proceedings of the 2nd International Conference on Applications of AI to Engineering*, Boston, MA, 1987.

M. E. French, *Conceptual Design for Engineers*, 2nd Edition, Design Council Books, London, 1985.

M. E. French, *Form, Structure and Mechanism*, MacMillan, London, 1992.

R. Ganeshan, *Reasoning with Design Intent*, PhD Dissertation, Department of Civil Engineering, Carnegie Mellon University, Pittsburgh, PA, May 1993.

R. Ganeshan, S. Finger, and J. H. Garrett, Jr., "Representing and Reasoning with Design Intent," in *Proceedings of the 1st International Conference on Artificial Intelligence in Design*, Edinburgh, June 1991.

J. H. Garrett, Jr., and S. J. Fenves, "A Knowledge-Based Standards Processor for Structural Component Design," *Engineering with Computers 2* (4), 1987.

J. H. Garrett, Jr., and S. J. Fenves, "Knowledge-Based Standard-Independent Member Design," *Journal of Structural Engineering 115* (6), 1989.

J. H. Garrett, Jr., and M. M. Hakim, "An Object-Oriented Model of Engineering Design Standards," *Journal of Computing in Civil Engineering, 6* (3), 1992.

L. Gasser and M. N. Huhns (Editors), *Distributed Artificial Intelligence*, Volume II, Morgan Kaufmann, San Mateo, CA, 1989.

J. S. Gero (Editor), *Design Optimization*, Academic Press, Orlando, FL, 1985.

J. S. Gero, "Prototypes: A New Schema for Knowledge-Based Design," Working Paper, Architectural Computing Unit, University of Sydney, Australia, 1987.

J. S. Gero (Editor), *Proceedings of AI in Design '92*, Kluwer Academic Publishers, Dordrecht, The Netherlands, 1992.

J. S. Gero and M. Yan, "Shape Emergence by Symbolic Reasoning," Working Paper, Design Computing Unit, University of Sydney, Australia, 1993.

R. Gronewold, C. White, C. Nichols, M. Opdahl, M. Shane, and K. Wong, *Design of a "Building Block" Analog Computer*, E4 Project Report, Department of Engineering, Harvey Mudd College, Claremont, CA, May 1993.

M. M. Hakim and J. H. Garrett, Jr., "Issues in Modeling and Processing Design Standards," *1992 Computers and Building Standards Workshop*, University of Montreal, Montreal, Canada, 1992.

B. Hartmann, B. Hulse, S. Jayaweera, A. Lamb, B. Massey, and R. Minneman, *Design of a "Building Block" Analog Computer*, E4 Project Report, Department of Engineering, Harvey Mudd College, Claremont, CA, May 1993.

S. I. Hayakawa, *Language in Thought and Action*, 4th Edition, Harcourt Brace Jovanovich, San Diego, CA, 1978.

B. Hayes-Roth, A. Garvey, V. Johnson, and M. Hewett, *A Modular and Layered Environment for Reasoning about Action*, Technical Report No. KSL 86-38, Department of Computer Science, Stanford University, Stanford, CA, April 1987.

M. R. Henderson, *Extraction of Feature Information from Three-Dimensional CAD Data*, PhD Dissertation, Department of Mechanical Engineering, Purdue University, West Lafayette, IN, May 1984.

M. R. Henderson and D. C. Anderson, "Computer Recognition and Extraction of Form Features: A CAD/CAM Link," *Computers in Industry, 6* (4), 1984.

D. Herbert, "Study Drawings in Architectural Design: Implications for CAD Systems," in *Proceedings of the 1987 Workshop of the Association for Computer-Aided Design in Architecture (ACADIA)*, 1987.

H. C. Howard and D. R. Rehak, "KADBASE: Interfacing Expert Systems with Databases," *IEEE Expert 4* (3), 1989.

D. Jain, G. P. Luth, H. Krawinkler, and K. H. Law, *A Formal Approach to Automating Conceptual Structural Design*, Technical Report No. 31, Center for Integrated Facility Engineering, Stanford University, Stanford, CA, 1990.

J. C. Jones, *Design Methods*, Wiley-Interscience, Chichester, UK, 1981.

D. Knuth, "Computer Science and Mathematics," *American Scientist, 61* (6), 1973.

R. E. Korf, "Search: A Survey of Recent Results," in H. E. Shrobe and AAAI (Editors), *Exploring Artificial Intelligence*, Morgan Kaufmann, San Mateo, CA, 1988.

F. L. Krause, F. H. Vosgerau, and N. Yaramanoglou, "Using Technical Rules and Features in Product Modelling," in *IFIP WG 5.2 First International Workshop on Intelligent CAD*, Cambridge, MA, October 1987.

J. E. Laird, A. Newell, and P. S. Rosenbloom, "Soar: An Architecture for General Intelligence," *Artificial Intelligence, 33* (1), 1987.

J. Larkin and H. A. Simon, "Why a Diagram Is (Sometimes) Worth a Thousand Words," *Cognitive Science, 11*, 1987.

J. K. Lathrop (Editor), *Life Safety Code Handbook*, 3rd Edition, National Fire Protection Association, Worcester, MA, 1985.

D. Lenat, "An Artificial Intelligence Approach to Discovery in Mathematics as Heuristic Search," in R. Davis and D. Lenat (Editors), *Knowledge-Based Systems in Artificial Intelligence*, McGraw-Hill, New York, 1982.

D. B. Lenat and E. A. Feigenbaum, "On the Thresholds of Knowledge," *Proceedings of IJCAI 1987*, Milan, Italy, August 1987.

R. E. Levitt, "Merging Artificial Intelligence with CAD: Intelligent, Model-Based Engineering," *Henry M. Shaw Memorial Lecture*, North Carolina State University, Raleigh, NC,

1990. Reprinted as Working Paper No. 28, Center for Integrated Facility Engineering, Stanford University, Stanford, CA, 1990.

R. E. Levitt, Y. Jin, and C. L. Dym, "Knowledge-Based Support for Management of Concurrent Multidisciplinary Design," *Artificial Intelligence for Engineering Design, Analysis and Manufacturing*, 5 (2), 1991.

E. C. Libardi, Jr., J. R. Dixon, and M. K. Simmons, "Designing with Features: Design and Analysis of Extrusions as an Example," Paper No. 86–DE–4, *ASME Design Engineering Conference*, Chicago, IL, March 1986.

K. Lien, G. Suzuki, and A. W. Westerberg, "The Role of Expert Systems Technology in Design," *Chemical Engineering Science*, 42 (5), 1987.

R. D. Logcher and R. E. Levitt, "Organization and Control of Engineering Design Firms," *ASCE Engineering Issues 105* (EI1), 1979.

S. C. Luby, J. R. Dixon, and M. K. Simmons, "Designing with Features: Creating and Using a Features Database for Evaluation of the Manufacturability of Castings," in *Proceedings of the ASME Computers in Engineering Conference*, ASME, Chicago, IL, July 1986.

R. D. Luce and H. Raiffa, *Games and Decisions*, John Wiley, New York, 1957.

M. P. Lukas and R. B. Pollock, "Automated Designs Through Artificial Intelligence Techniques," *The AI and Advanced Computer Technology Conference*, Long Beach, CA, 4–6 May 1988.

G. P. Luth, *Reasoning and Representation for Integrated Structural Design of High-Rise Commercial Office Buildings*, Ph.D. Dissertation, Department of Civil Engineering, Stanford University, Stanford, CA, 1990.

M. L. Maher, *HI-RISE: A Knowledge-Based Expert System for the Preliminary Design of High-Rise Buildings*, Ph.D. Dissertation, Department of Civil Engineering, Carnegie Mellon University, Pittsburgh, PA, 1984.

M. L. Maher and S. J. Fenves, "HI-RISE: An Expert System for the Preliminary Structural Design of High-Rise Buildings," in J. S. Gero (Editor), *Knowledge Engineering in Computer-Aided Design*, North-Holland, Amsterdam, 1985.

M. L. Maher and P. Longinos, "Development of an Expert System Shell for Engineering Design," *International Journal of Applied Engineering Education*, 3 (3), 1987.

J. McDermott, "R1: The Formative Years," *AI Magazine*, 2 (2), 1981.

J. McDermott, "R1: A Rule-Based Configuration of Computer Systems," *Artificial Intelligence*, 19 (1), 1982.

F. C. Mish (Editor in Chief), *Webster's Ninth New Collegiate Dictionary*, Merriam Webster, Springfield, MA, 1983.

S. Mittal and A. Araya, "A Knowledge-Based Framework for Design," in *Proceedings of AAAI-86*, AAAI, Philadelphia, PA, 1986.

S. Mittal, D. G. Bobrow, and J. de Kleer, *DARN: A Community Memory for a Diagnosis and Repair Task*, Technical Memorandum, Xerox Palo Alto Research Center, Palo Alto, CA, 1984.

S. Mittal and C. L. Dym, "Knowledge Acquisition from Multiple Experts," *AI Magazine*, 6 (2), 1985.

S. Mittal, C. L. Dym, and M. Morjaria, "PRIDE: An Expert System for the Design of Paper Handling Systems," *IEEE Computer*, 19 (7), 1986.

A. K. Modi, A. Newell, D. M. Steier, and A. W. Westerberg, "Building a Chemical Process Design System within SOAR: Part I–Design Issues; Part II–Learning Issues," unpublished manuscript (submitted for publication in *Journal of Computers and Chemical Engineering*), 1992.

M. Morjaria, "Knowledge-Based Systems or Engineering Design," in *AUTOFACT '89 Conference Proceedings*, Detroit, MI, 1989.

M. Morjaria, S. Mittal, and C. L. Dym, "Interactive Graphics in Expert Systems for Engineering Applications," in *Proceedings of the 1985 International Computers in Engineering Conference*, Boston, MA, August 1985.

J. Mostow, "Towards Better Models of the Design Process," *AI Magazine*, 6 (1), 1985.

T. Murakami and N. Nakajima, "Research on Design-Diagnosis Using Feature Description," in *IFIP WG 5.2 First International Workshop on Intelligent CAD*, Cambridge, MA, October 1987.

A. Newell, "The Knowledge Level," *AI Magazine, 2* (2), 1981.

A. Newell and H. A. Simon, "GPS: A Program that Simulates Human Thought," in E. A. Feigenbaum and J. Feldman (Editors), *Computers and Thought*, McGraw-Hill, New York, 1963.

A. Newell and H. A. Simon, *Human Problem Solving*, Prentice-Hall, Englewood Cliffs, NJ, 1972.

H. P. Nii, "Blackboard Systems: Part I and Part II," *AI Magazine, 7* (2, 3), 1986.

H. Nii, "Blackboard Systems," in A. Barr, P. R. Cohen, and E. A. Feigenbaum (Editors), *Handbook of Artificial Intelligence*, 4, Addison-Wesley, Reading, MA, 1980.

N. J. Nilsson, *Principles of Artificial Intelligence*, Morgan Kaufmann, Los Altos, CA, 1993.

NTIS, *PDES/STEP Standard*, National Institute of Standards and Technology, NITS–IR–88–4004, Gaithersburg, MD, 1989.

K. N. Otto and E. K. Antonsson, "Trade-Off Strategies in Engineering Design," *Research in Engineering Design, 3*, 1991.

G. Pahl and W. Beitz, *Engineering Design*, Design Council Books, London, 1984.

A. Palladio, *The Four Books of Architecture*, Dover, New York, 1965.

P. Y. Papalambros and D. J. Wilde, *Principles of Optimal Design*, Cambridge University Press, Cambridge, 1988.

H. Petroski, *To Engineer Is Human*, St. Martin's Press, New York, 1982.

E. O. Pfrang, "Collapse of the Kansas City Hyatt Regency Walkways," *Civil Engineering, 52* (7), 1982.

M. J. Pratt, "Solid Modeling and the Interface between Design and Manufacture," *IEEE Computer Graphics and Applications, 7* (4), July 1984.

M. J. Pratt and P. H. Wilson, *Requirements for Support of Form Features in a Solid Modeling System*, Final Report R–85–ASPP–01, Computer-Aided Manufacturing-International (CAM-I), Arlington, TX, 1985.

P. Pu (Guest Editor), "Special Issue on Case-Based Reasoning in Design," *Artificial Intelligence for Engineering Design, Analysis and Manufacturing, 7* (2), 1993.

D. R. Rehak, Personal communication, 10 August 1988.

D. R. Rehak, Personal communication, 31 August 1992.

E. Rich, *Artificial Intelligence*, McGraw-Hill, New York, 1983.

J. R. Rinderle, "Function, Form, Fabrication: Considerations in Design Automation," Presentation at the Workshop on Machine Intelligence in Machine Design, Industrial Technology Institute, Ann Arbor, MI, 1985.

J. R. Rinderle et al., "Form-Function Characteristics of Mechanical Designs—Research in Progress," prepared for the 1988 NSF Grantee Workshop on Design Theory and Methodology, Rensselaer Polytechnic Institute, Troy, NY, 1988.

M. A. Rosenman and J. S. Gero, "Design Codes as Expert Systems," *Computer-Aided Design, 17* (9), 1985.

J. Runkel, W. P. Birmingham, T. Darr, B. Maxim, and I. D. Tommelein, "Domain-Independent Design System: Environment for Rapid Development of Configuration Design Systems," in J. S. Gero (Editor), *Proceedings of AI in Design '92*, Kluwer Academic Publishers, Dordrecht, The Netherlands, 1992.

M. D. Rychener (Editor), *Expert Systems for Engineering Design*, Academic Press, Boston, 1988.

S. E. Salata and C. L. Dym, "Representing Strategic Choices in Structural Modeling," *Journal of Computing in Civil Engineering, 5* (4), October 1991.

M. Salvadori, *Why Buildings Stand Up*, McGraw-Hill, New York, 1980.

G. Schmitt, "ARCHPLAN – An Architectural Planning Front End to Engineering Design Expert Systems," in M. D. Rychener (Editor), *Expert Systems for Engineering Design*, Academic Press, Boston, MA, 1988.

D. A. Schon, *The Reflective Practitioner*, Basic Books, New York, 1983.

H. A. Simon, "Style in Design," in C. M. Eastman (Editor), *Spatial Synthesis in Computer-Aided Building Design*, Applied Science Publishers, London, England, 1975.

H. A. Simon, *The Sciences of the Artificial*, 2nd edition, MIT Press, Cambridge, MA, 1981.

D. Sriram, R. Logcher, and S. Fukuda (Editors), in *Proceedings of the MIT-JSME Workshop on Cooperative Product Development*, MIT Press, Cambridge, MA, November 1989.

F. I. Stahl, R. N. Wright, S. J. Fenves, and J. R. Harris, "Expressing Standards for Computer-Aided Building Design," *Computer-Aided Design*, *15* (6), 1983.

L. Stauffer, *An Empirical Study on the Process of Mechanical Design*, Thesis, Department of Mechanical Engineering, Oregon State University, Corvallis, OR, 1987.

L. Stauffer, D. G. Ullman, and T. G. Dietterich, "Protocol Analysis of Mechanical Engineering Design," in *Proceedings of the 1987 International Conference on Engineering Design*, Boston, MA, 1987.

L. Steels, "Components of Expertise," *AI Magazine*, *11* (2), 1990.

D. M. Steier, Personal communication, 9 March 1993.

M. J. Stefik, "The Knowledge Medium," *AI Magazine*, *7* (1), 1986.

M. J. Stefik, *Introduction to Knowledge Systems*, Xerox Palo Alto Research Center, Palo Alto, CA, 1990.

M. J. Stefik, J. Aikins, R. Balzer, J. Benoit, L. Birnbaum, F. Hayes-Roth, and E. Sacerdoti, *The Organization of Expert Systems: A Prescriptive Tutorial*, Technical Report No. VLSI-82-1, Xerox Palo Alto Research Center, Palo Alto, CA, 1982.

M. J. Stefik and D. G. Bobrow, "Object-Oriented Programming: Themes and Variations," *AI Magazine*, *6* (4), 1985.

M. J. Stefik and L. Conway, "The Principled Engineering of Knowledge," *AI Magazine*, *3* (3), 1982.

L. Steinberg and R. Ling, *A Priori Knowledge of Structure vs. Constraint Propagation: One Fragment of a Science of Design*, Working Paper 164, Rutgers AI/Design Group, Rutgers University, New Brunswick, NJ, 1990.

G. Stevens, *The Reasoning Architect: Mathematics and Science Design*, McGraw-Hill, New York, 1990.

G. Stiny, "An Introduction to Shape and Shape Grammars," *Planning and Design: Environment and Planning B*, *7*, 1980.

G. Stiny, "Environments, Languages, Representatives, and Data Bases for Design," Lecture at the 1988 NSF Grantee Workshop on Design Theory and Methodology, Rensselaer Polytechnic Institute, Troy, NY, 1988.

G. Stiny, Personal communication, 3 June 1988.

G. Stiny and J. Gips, *Algorithmic Aesthetics*, University of California Press, Berkeley, CA, 1978.

M. Stonebraker and L. Rowe, "The Design of POSTGRES," in *Proceedings of the ACM SIGMOD Conference*, 1986.

J. C. Tang, *A Framework for Understanding the Workspace Activity for Design Teams*, Technical Report P88–00074, Xerox Palo Alto Research Center, Palo Alto, CA, 1988.

J. Thompson, *Organizations in Action*, McGraw-Hill, New York, 1967.

C. Tong and D. Sriram (Editors), *Artificial Intelligence in Engineering Design, Volume I: Design Representation and Models of Routine Design*, Academic Press, Boston, MA, 1992a.

C. Tong and D. Sriram (Editors), *Artificial Intelligence in Engineering Design, Volume II: Models of Innovative Design, Reasoning about Physical Systems, and Reasoning about Geometry*, Academic Press, Boston, MA, 1992b.

C. Tong and D. Sriram (Editors), *Artificial Intelligence in Engineering Design, Volume III: Knowledge Acquisition, Commercial Applications and Integrated Environments*, Academic Press, Boston, MA, 1992c.

D. S. Touretzky, *The Mathematics of Inheritance Systems*, Pitman, London, and Morgan Kaufmann, Los Altos, CA, 1986.

G. M. Turkiyyah and S. J. Fenves, "Knowledge-Based Analysis of Structural Systems," in *Proceedings of the Second International Conference on Artificial Intelligence in Engineering*, II, Boston, MA, 1987.

UBC, *Uniform Building Code*, International Conference of Building Officials, Whittier, CA, 1988.

D. G. Ullman, "A Taxonomy for Mechanical Design," *Research in Engineering Design*, 3, 1992a.

D. G. Ullman, *The Mechanical Design Process*, McGraw-Hill, New York, 1992b.

D. G. Ullman and T. G. Dietterich, "Toward Expert CAD," *Computers in Mechanical Engineering*, 6 (3), 1987.

D. G. Ullman, T. G. Dietterich, and L. Stauffer, "A Model of the Mechanical Design Process Based on Empirical Data," *Artificial Intelligence for Engineering Design, Analysis and Manufacturing*, 2 (1), 1988.

D. G. Ullman, S. Wood, and D. Craig, "The Importance of Drawing in the Mechanical Design Process," *Computers and Graphics*, 14 (2), 1990.

M. Vaghul, J. R. Dixon, G. E. Zinsmeister, and M. K. Simmons, "Expert Systems in a CAD Environment: Injection Molding Part Design as an Example," *Proceedings of the ASME Computers in Engineering Conference*, Boston, MA, August 1985.

G. N. Vanderplaats, *Numerical Optimization Techniques for Engineering Design*, McGraw-Hill, New York, 1984.

VDI, *VDI–2221: Systematic Approach to the Design of Technical Systems and Products*, Verein Deutscher Ingenieure, VDI-Verlag, Translation of the German edition 11/1986, 1987.

J. Walton, *Engineering Design: From Art to Practice*, West Publishing, St. Paul, MN, 1991.

M. A. Wesley, T. Lozano-Perez, L. I. Lieberman, M. A. Lavin, and D. D. Grossman, "A Geometric Modeling System for Automated Mechanical Assembly," *IBM Journal of Research and Development*, 24 (1), 1980.

D. J. Wilde, *Globally Optimal Design*, John Wiley, New York, 1978.

D. J. Wilde, "Analysis to Support Design," Lecture at the 1988 NSF Grantee Workshop on Design Theory and Methodology, Rensselaer Polytechnic Institute, Troy, NY, 1988.

T. Winograd and F. Flores, *Understanding Computers and Cognition*, Ablex Publishing, Norwood, NJ, 1986.

P. H. Winston, *Artificial Intelligence*, 3rd Edition, Addison-Wesley, Reading, MA, 1993.

T. C. Woo, "Interfacing Solid Modeling to CAD and CAM: Data Structures and Algorithms for Decomposing a Solid," *Computer Integrated Manufacturing*, ASME Production Engineering Division, New York, November 1983.

K. L. Wood and E. K. Antonsson, "Computations with Imprecise Parameters in Engineering Design: Background and Theory," *Journal of Mechanisms, Transmissions and Automation in Design*, 111, 1989.

K. L. Wood and E. K. Antonsson, "Modeling Imprecision and Uncertainty in Preliminary Engineering Design," *Mechanical Machine Theory*, 25 (3), 1990.

T. T. Woodson, *Introduction to Engineering Design*, McGraw-Hill, New York, 1966.

H. H. Woolf (Editor), *Webster's New Collegiate Dictionary*, Merriam Webster, Springfield, MA, 1977.

R. N. Wright, S. J. Fenves, and J. R. Harris, *Modeling of Standards: Technical Aids for Their Formulation, Expression and Use*, National Bureau of Standards, Washington, DC, March 1980.

C. Zozaya-Gorostiza, C. Hendrickson, and D. R. Rehak, *Knowledge-Based Process Planning for Construction and Manufacturing*, Academic Press, Boston, MA, 1989.

New References

S. Ahmed, S. Kim, and K. M. Wallace, "A Methodology for Creating Ontologies for Engineering Design," *Journal of Computing and Information Science in Engineering, 7* (2), 2007.

U. Athavankar, P. Bokil, K. Guruprasad, R. Patsute, and S. Sharma, "Reaching Out in the Mind's Space," *Design Computing and Cognition '08*, J. S. Gero and A. K. Goel (Eds.), Springer, 2008.

M. Aurisicchio, R. H. Bracewell, and K. M. Wallace, "Characterising in Detail the Information Requests of Engineering Designers," *Proceeding of the ASME International Design Engineering Technical Conferences*, DETC2006-99418, 2006.

F. Baader, D. Calvanese, D. McGuinness, D. Nardi, and P. Patel-Schneider (Eds.), *The Description Logic Handbook: Theory, Implementation and Applications*, Cambridge University Press, 2003.

M. E. Balazs and D. C. Brown, "Design Simplification by Analogical Reasoning," in *Knowledge Intensive CAD to Knowledge Intensive Engineering*, U. Cugini and M. Wozny (Eds.), Kluwer Academic, 2002.

Z. Bilda, J. S. Gero, and T. Purcell, "To Sketch or Not to Sketch? That Is the Question," *Design Studies, 27* (5), 2006.

W. P. Birmingham, A. P. Gupta, and D. P. Siewiorek, *Automating the Design of Computer Systems: The MICON Project*, Jones and Bartlett, 1992.

D. T. Bischel, T. Stahovich, E. Peterson, R. Davis, and A. Adler, "Combining Speech and Sketch to Interpret Unconstrained Descriptions of Mechanical Devices," *Proceeding of the International Joint Conference on AI*, 2009.

M. A. Boden, "What Is Creativity?" *Dimensions of Creativity*, M. A. Boden (Ed.), The MIT Press, 1994.

M. Bohm, R. Stone, and S. Szykman, "Enhancing Virtual Product Representations for Advanced Design Repository Systems," *Journal of Computer and Information Science in Engineering, 5* (4), 2005.

S. Borgo, M. Carrara, P. Garbacz, and P. E. Vermaas, "A Formal Ontological Perspective on the Behaviors and Functions of Technical Artifacts," *Developing and Using Engineering Ontologies*, Special Issue, C. McMahon and J. van Leeuwen (Eds.), *AIEDAM, 23* (1), 2009.

R. J. Brachman and H. R. Levesque, *Knowledge Representation and Reasoning*, Morgan Kaufmann, 2004.

B. Bredeweg and P. Struss, "Current Topics in Qualitative Reasoning," *AI Magazine*, AAAI, Winter 2003. Available at http://staff.science.uva.nl/~bredewegpdfaimag2003a.pdf

D. C. Brown, "Failure Handling in a Design Expert System," *Computer-Aided Design, 17* (9), 1985.

D. C. Brown, "Compilation: The Hidden Dimension of Design Systems," *Intelligent CAD, III*, H. Yoshikawa, F. Arbab, and T. Tomiyama (Eds.), IFIP/North-Holland, 1991.

D. C. Brown, "Design," *Encyclopedia of Artificial Intelligence, 2nd edition*, S. C. Shapiro (Ed.), John Wiley, 1992.

D. C. Brown, "Routineness Revisited," *Mechanical Design: Theory and Methodology*, M. Waldron and K. Waldron (Eds.), Springer-Verlag, 1996.

D. C. Brown, "Defining Configuring," *Configuration*, Special Issue, T. Darr, D. McGuinness, and M. Klein (Eds.), *AIEDAM, 12* (4), 1998.

D. C. Brown, "Functional, Behavioral and Structural Features," *Proceeding of the Design Theory and Methodology Conference*, ASME Design Technical Conferences, Chicago, IL, 2003.

D. C. Brown, "Assumptions in Design and Design Rationale," *Proceeding of the Design Rationale Workshop*, DCC'06: Design Computing and Cognition Conference, Eindhoven, The Netherlands, 2006.

D. C. Brown, "Guiding Computational Design Creativity Research," *Studying Design Creativity*, J. S. Gero (Ed.), Springer, 2012. Available at http://web.cs.wpi.edu/~dcb/Papers/sdc08-paper-Brown-25-Feb.pdf

D. C. Brown (Ed.), "Problem Solving Methods: Past, Present and Future," Special Issue, *AIEDAM, 23* (3), 2009.

D. C. Brown and B. Chandrasekaran, *Design Problem Solving*, Pitman, London, and Morgan Kaufmann, Los Altos, CA, 1989.

K. N. Brown, "Grammatical Design," Special Issue on Artificial Intelligence in Design, *IEEE Expert: Intelligent Systems and Their Applications, 12* (2), 1997.

J. E. Burge and R. H. Bracewell, "Design Rationale," Special Issue, *AIEDAM, 22* (4), 2008.

J. E. Burge and D. C. Brown, *Design Rationale Types and Tools*, Technical Report, Fall 1998. Available at http://web.cs.wpi.edu/Research/aidg/DR-Rpt98.html

J. E. Burge, J. M. Carroll, R. McCall, and I. Mistrík, *Rationale-Based Software Engineering*, Springer-Verlag, 2008.

J. Cagan, "Engineering Shape Grammars: Where We Have Been and Where We Are Going," *Formal Engineering Design Synthesis*, E. A. Antonsson and J. Cagan (Eds.), Cambridge University Press, 2001.

J. Cagan, M. I. Campbell, S. Finger, and T. Tomiyama, "A Framework for Computational Design Synthesis: Models and Applications," *Journal of Computing and Information Science in Engineering, 5* (3), 2005.

M. Campbell, J. Cagan, and K. Kotovsky, "A-Design: An Agent-Based Approach to Conceptual Design in a Dynamic Environment," *Research in Engineering Design, 11* (3), 1999.

A. Chakrabarti, K. Shea, R. Stone, M. Campbell, J. Cagan, N. Vargas-Hernandez, and K. L. Wood, "Computer-Based Design Synthesis Research: An Overview," *Journal of Computing and Information Science in Engineering, 11* (2), 2011.

A. Chakrabarti and L. Shu (Eds.), "Biologically Inspired Design," Special Issue, *AIEDAM, 24* (4), 2010.

B. Chandrasekaran, "Design Problem Solving: A Task Analysis," Special Issue on Design, J. S. Gero and M. L. Maher (Eds.), *AI Magazine, 11* (4), 1990.

B. Chandrasekaran and T. R. Johnson, "Generic Tasks and Task Structures: History, Critique and New Directions," *Second Generation Expert Systems*, J. M. David, J. P. Krivine, and R. Simmons (Eds.), Springer-Verlag, 1993.

B. Chandrasekaran and J. R. Josephson, "Function in Device Representation," Special Issue on Computer Aided Engineering, *Engineering with Computers, 16*, 2000.

E. Charniak, "Bayesian Networks without Tears," *AI Magazine, 12* (4), 1991.

W. J. Clancey, "Heuristic Classification," *Artificial Intelligence, 27* (3), 1985.

N. Cross, "Design Cognition: Results from Protocol and Other Empirical Studies of Design Activity," *Design Knowing and Learning: Cognition in Design Education*, C. Eastman, W. Newstetter, and M. McCracken (Eds.), Elsevier, 2001.

N. Cross, H. Christiaans, and K. Dorst, *Analysing Design Activity*, John Wiley, 1996.

M. Danesh and Y. Jin, "An Agent-Based Decision Network for Concurrent Engineering," *Journal of Concurrent Engineering Research and Applications*, *9* (1), 2001.

DBD, *Decision Based Design Open Workshop*, 2004. Available at http://dbd.eng.buffalo.edu

R. de Neufville, Uncertainty Management for Engineering Systems Planning and Design, Engineering Systems Monograph, MIT Engineering Systems Division, Cambridge, MA, 2004. Available at http://esd.mit.edu/symposium/pdfs/monograph/uncertainty.pdf

R. Dechter, *Constraint Processing*, Morgan Kaufmann, 2003.

J. R. Dixon, M. K. Simmons, and P. R. Cohen, "An Architecture for the Application of Artificial Intelligence to Design," *Proceeding of the ACM/IEEE 21st Annual Design Automation Conference*, IEEE, Albuquerque, NM, 1984.

P. E. Doepker and C. L. Dym (Eds.), "Design Engineering Education," Special Issue, *Journal of Mechanical Design*, *129* (7), 2007.

A. H. B. Duffy, "The 'What' and 'How' of Learning in Design," Special Issue on Artificial Intelligence in Design, *IEEE Expert: Intelligent Systems and Their Applications*, *12* (3), 1997.

A. H. B. Duffy and F. M. T. Brazier (Eds.), "Learning and Creativity in Design," Special Issue, *AIEDAM*, *18* (3/4), 2004.

C. L. Dym, "Learning Engineering: Design, Languages, and Experiences," *Journal of Engineering Education*, *88* (2), 1999.

C. L. Dym, *Principles of Mathematical Modeling*, *2nd Edition*, Elsevier Academic Press, 2004.

C. L. Dym, A. M. Agogino, O. Eris, D. D. Frey, and L. J. Leifer, "Engineering Design Thinking, Teaching, and Learning," *Journal of Engineering Education*, *94*, 2005.

C. L. Dym and P. Little, with E. J. Orwin and R. E. Spjut, *Engineering Design: A Project Based Introduction, 3rd Edition*, John Wiley, 2009.

C. M. Eastman, Y.-S. Jeong, R. Sacks, and I. Kaner, "Exchange Model and Exchange Object Concepts for Implementation of National BIM Standards," *Journal of Computing in Civil Engineering*, *24* (1), 2010.

C. Eastman, P. Teicholz, R. Sacks, and K. Liston, *BIM Handbook: A Guide to Building Information Modeling for Owners, Managers, Designers, Engineers, and Contractors*, 2nd Edition, John Wiley, 2011.

M. S. Erden, H. Komoto, T. J. van Beek, V. D'Amelio, E. Echavarria, and T. Tomiyama, "A Review of Function Modeling: Approaches and Applications," *AIEDAM*, *22* (2), 2008.

K. Ericsson and H. Simon, *Protocol Analysis: Verbal Reports as Data, 2nd Edition*, MIT Press, 1993.

A. Felfernig, M. Stumptner, and J. Tiihonen, "Configuration," Special Issue, *AIEDAM*, *25* (2), 2011.

P. J. Feltovich, K. M. Ford, and R. R. Hoffman (Eds.), *Expertise in Context: Human and Machine*, MIT Press, 1997.

D. Fensel, *Problem-Solving Methods: Understanding, Description, Development, and Reuse*, Lecture Notes in AI (LNAI) 1791, Springer, 2000.

R. Finkel and B. O'Sullivan, "Reasoning about Conditional Constraint Specification Problems and Feature Models," *AIEDAM*, *25* (2), 2011.

D. D. Frey, W. P. Birmingham, and C.L. Dym (Eds.), "Design Pedagogy: Representations and Processes," Special Issue, *AIEDAM*, *24* (3), 2010.

R. Fruchter and M. L. Maher, "Support for Design Teams," Special Issue, *AIEDAM*, *21* (3), 2007.

L. Fua and L. B. Kara, "From Engineering Diagrams to Engineering Models: Visual Recognition and Applications," *Computer-Aided Design*, *43* (3), 2011.

J. S. Gero, "Design Prototypes: A Knowledge Representation Schema for Design," Special Issue on Design, J. S. Gero and M. L. Maher (Eds.), AAAI, *AI Magazine*, *11* (4), 1990.

J. S. Gero, J. W. T. Kan, and M. Pourmohamadi, "Analyzing Design Protocols: Development of Methods and Tools," *Research into Design – Supporting Sustainable Product Development*, A. Chakrabarti (Ed.), Research Publishing Services, 2011.

J. S. Gero and U. Kannengiesser, "An Ontology of Situated Design Teams," *AIEDAM, 21* (3), 2007.

J. S. Gero and U. Kannengiesser, "An Ontological Account of Donald Schön's Reflection in Designing," *International Journal of Design Sciences and Technology, 15* (2), 2008.

J. S. Gero and M. L. Maher (Eds.), *Modeling Creativity and Knowledge-Based Creative Design*, Lawrence Erlbaum, 1993.

M. Ghallab, D. Nau, and P. Traverso, *Automated Planning: Theory and Practice*, Morgan Kaufmann, 2004.

A. K. Goel, "Design, Analogy, and Creativity," *IEEE Expert: Intelligent Systems and Their Applications, 12* (3), 1997.

A. Goel, S. Bhatta, and E. Stroulia, "Kritik: An Early Case-Based Design System," *Issues and Applications of Case-Based Reasoning in Design*, M. L. Maher and P. Pu (Eds.), Erlbaum, 1997.

A. Goel, S. Rugaber, and S. Vattam, "Structure, Behavior, and Function of Complex Systems: The Structure, Behavior, and Function Modeling Language," *Developing and Using Engineering Ontologies*, Special Issue, C. McMahon and J. van Leeuwen (Eds.), *AIEDAM, 23* (1), 2009.

D. E. Goldberg, *Genetic Algorithms in Search, Optimization, and Machine Learning*, Addison-Wesley, 1989.

A. Gómez-Pérez, M. Fernandez-Lopez, and O. Corcho, *Ontological Engineering: With Examples from the Areas of Knowledge Management, e-Commerce and the Semantic Web*, Springer, 2004.

D. Grecu and D. C. Brown, "Dimensions of Learning in Design," Machine Learning in Design, Special Issue, A. H. B. Duffy, D. C. Brown, and A. K. Goel (Eds.), *AIEDAM, 12*, 1998.

G. R. Greenfield, "On the Origins of the Term 'Computational Aesthetics'," *Proceeding of the 1st EG Workshop on Computational Aesthetics in Graphics, Visualization and Imaging*, Spain, 2005.

K. Gurney, *An Introduction to Neural Networks*, CRC Press, 1997.

K. J. Hammond, *Case-Based Planning: Viewing Planning as a Memory Task*. Academic Press, 1989.

S. Harris, Science Cartoons Plus: The Cartoons of S. Harris. 2011. Available at http://www.sciencecartoonsplus.com

A. Haug, "The Illusion of Tacit Knowledge as the Great Problem in the Development of Product Configurators," *AIEDAM, 26* (1), 2012.

J. Hendler, A. Tate, and M. Drummond, "AI Planning: Systems and Techniques," *AAAI, AI Magazine, 11* (2), 1990. Available at http://www.aaai.org/AITopics/assets/PDF/AIMag11-02-006.pdf

A. Heylighen, "Studying the Unthinkable Designer: Designing in the Absence of Sight," *Design Computing and Cognition '10*, J. S. Gero (Ed.), Springer, 2011.

J. Hirtz, R. B. Stone, S. Szykman, D. A. McAdams, and K. L. Wood, "A Functional Basis for Engineering Design: Reconciling and Evolving Previous Efforts," *Journal of Research in Engineering Design, 13* (2), 2002.

P. Jackson, *Introduction to Expert Systems*, 3rd Edition, Addison-Wesley, 1999.

M. J. Jakiela, C. Chapman, J. Duda, A. Adewuya, and K. Saitou, "Continuum Structural Topology Design with Genetic Algorithms," *Computer Methods in Applied Mechanics and Engineering, 186*, (2–4), 2000.

T. A. W. Jarratt, C. M. Eckert, N. H. M. Caldwell, and P. J. Clarkson, "Engineering Change: An Overview and Perspective on the Literature," *Research in Engineering Design, 22* (2), 2011.

L. Joskowicz and E. Sacks, "Computer-Aided Mechanical Assembly Design Using Configuration Spaces," Purdue University Computer Science Technical Reports, Report No. 97-001, 1997. Available at http://docs.lib.purdue.edu/cstech/1341

E. Kant, "Understanding and Automating Algorithm Design," *IEEE Transactions on Software Engineering, SE-11* (11), 1985.

L. B. Kara and M. C. Yang, "Sketching, and Pen-Based Design Interaction," Special Issue, *AIEDAM*, *26* (3), 2012.

R. Kicinger, T. Arciszewski, and K. A. De Jong, "Evolutionary Computation and Structural Design: A Survey of the State of the Art," *Computers and Structures*, *83* (23/24), 2005.

M. Klein, "Supporting Conflict Resolution in Cooperative Design Systems," *IEEE Transactions on Systems, Man and Cybernetics*, *21* (6), 1991.

M. Kleinsmann and A. Maier (Eds.), "Studying and Supporting Design Communication," Special Issue, *AIEDAM*, *27* (2), 2013.

J. R. Koza, M. A. Keane, M. J. Streeter, T. P. Adams, and L. W. Jones, "Invention and Creativity in Automated Design by Means of Genetic Programming," *AIEDAM*, *18* (3), 2004.

T. Kurtoglu, A. Swantner, and M. I. Campbell, "Automating the Conceptual Design Process: From Black Box to Component Selection," *AIEDAM*, *24* (1), 2010.

Y. Labrou, T. Finin, and Y. Peng, "Agent Communication Languages: The Current Landscape," *IEEE Intelligent Systems Magazine*, *14*, 1999.

J. E. Laird, A. Newell, and P. S. Rosenbloom, "SOAR: An Architecture for General Intelligence," *The Soar Papers: Research in Integrated Intelligence*, P. Rosenbloom, J. Laird, and A. Newell (Eds.), Vol. 1, MIT Press, 1993.

S. E. Lander, "Issues in Multiagent Design Systems," *IEEE Expert: Intelligent Systems and Their Applications*, *12* (2), 1997.

G. Lee, C. M. Eastman, T. Taunk, and C.-H. Ho, "Usability Principles and Best Practices for the User Interface Design of Complex 3D Architectural Design and Engineering Tools," *International Journal of Human-Computer Studies*, *68* (1/2), 2010.

J. Liu and D. C. Brown, "Generating Design Decomposition Knowledge for Parametric Design Problems," *Artificial Intelligence in Design '94*, J. S. Gero and F. Sudweeks (Eds.), Kluwer Academic, 1994.

M. L. Maher, M. B. Balachandran, and D. M. Zhang, *Case-Based Reasoning in Design*, Lawrence Erlbaum, 1995.

M. L. Maher and P. Pu, *Issues and Applications of Case-Based Reasoning in Design*, Lawrence Erlbaum, 1997.

S. Marcus, J. Stout, and J. McDermott, "VT: An Expert Elevator Designer That Uses Knowledge-Based Backtracking," *AI Magazine*, *8* (4), 1987.

D. A. McAdams and C. L. Dym, "Modeling and Information in the Design Process," *Proc. ASME International Design Engineering Technical Conferences*, DETC2004-57101, 2004.

J. McDermott, "R1: A Rule-Based Configurer of Computer Systems," *Artificial Intelligence*, *19*, North-Holland, 1982.

A. F. McKenna and A. R. Carberry, "Characterizing the Role of Modeling in Innovation," *International Journal of Engineering Education*, *28* (2), 2012.

MDW, *Mudd Design Workshops*, n.d. Available at http://www.hmc.edu/academics clinicresearch/interdisciplinarycenters/cde1/workshops.html

M. Mitchell, *An Introduction to Genetic Algorithms*, MIT Press, 1998.

S. Mittal and C. L. Dym, "Knowledge Acquisition from Multiple Experts," *AI Magazine*, *6* (2), 1985.

S. Mittal, C. L. Dym, and M. Morjaria, "PRIDE: An Expert System for the Design of Paper Handling Systems," *IEEE Computer*, *19* (7), 1986.

E. Motta, *Reusable Components for Knowledge Modeling: Case Studies in Parametric Design Problem Solving*, IOS Press, 1999.

T. Murakami, "Retrieval Using Configuration Spaces," *Engineering Design Synthesis: Understanding, Approaches and Tools*, A. Chakrabarti (Ed.), Springer-Verlag, 2002.

S. Narasimhan, K. P. Sycara, and D. Navin-Chandra, "Representation and Synthesis of Non-Monotonic Mechanical Devices," *Issues and Applications of Case-Based Reasoning in Design*, M. L. Maher and P. Pu (Eds.), Lawrence Erlbaum, 1997.

P. P. Nayak, L. Joskowicz, and S. Addanki, "Automated Model Selection Using Context-Dependent Behaviors," *Proceeding of the 10th National Conference on AI*, AAAI Press, 1991.

N. F. Noy and D. L. McGuinness, "Ontology Development 101: A Guide to Creating Your First Ontology," Stanford Knowledge Systems Laboratory, Technical Report KSL-01-05 and Stanford Medical Informatics Technical Report SMI-2001-0880, 2001. Available at http://protege.stanford.edu/publications/ontology_development/ontology101-noy-mcguinness.html

Y. Oh, M. D. Gross, and E. Y.-L. Do, "Computer-Aided Critiquing Systems: Lessons Learned and New Research Directions," *Proceeding of the Computer Aided Architectural Design Research in Asia* (CAADRIA), Thailand, 2008.

B. O'Sullivan (Ed.), "Constraints and Design," Special Issue, *AIEDAM*, *20* (4), 2006.

A. Pease, *Suggested Upper Merged Ontology (SUMO)*, 2011. Available at http://www.ontologyportal.org

H. Petroski, *To Engineer Is Human: The Role of Failure in Successful Design*, Vintage Books, 1992.

J. B. Pollack, G. S. Hornby, H. Lipson, and P. Funes, "Computer Creativity in the Automatic Design of Robots," *Leonardo, Journal for the International Society for Arts Sciences and Technology*, *36* (2), 2003.

W. Regli, J. Kopena, M. Grauer, T. Simpson, R. Stone, K. Lewis, M. Bohm, D. Wilkie, M. Piecyk, and J. Osecki, "Archiving the Semantics of Digital Engineering Artifacts: A Case Study," *AI Magazine*, *31*, 2010.

S. Russell and P. Norvig, *Artificial Intelligence: A Modern Approach*, 3rd Edition, Prentice Hall, 2010.

T. A. Salomone, *What Every Engineer Should Know about Concurrent Engineering*, CRC Press, 1995.

D. A. Schön, *The Reflective Practitioner: How Professionals Think in Action*, Basic Books, 1983.

J. J. Shah and M. Mäntylä, *Parametric and Feature-Based CAD/CAM: Concepts, Techniques, and Applications*, John Wiley, 1995.

J. J. Shah, S. M. Smith, N. Vargas-Hernandez, D. R. Gerkens, and M. Wulan, "Empirical Studies of Design Ideation: Alignment of Design Experiments with Lab Experiments," *Proceeding of the ASME International Design Engineering Technical Conferences*, DETC2003/DTM-48679, 2003a.

J. J. Shah, N. Vargas-Hernandez, and S. M. Smith, "Metrics for Measuring Ideation Effectiveness," *Design Studies*, *24*, 2003b.

W. Shen, D. H. Norrie, and J. P. Barthes, *Multi-Agent Systems for Concurrent Intelligent Design and Manufacturing*, Taylor & Francis, 2001.

B. G. Silverman, "Survey of Expert Critiquing Systems: Practical and Theoretical Frontiers," *Communications of the ACM*, *35* (4), 1992.

T. W. Simpson, "Product Platform Design and Customization: Status and Promise," *AIEDAM*, *18* (1), 2005.

K. Sims, "Evolving Virtual Creatures," *Proceeding of the ACM SIGGRAPH*, 1994.

T. Smithers, "Towards a Knowledge Level Theory of Design Process," *Artificial Intelligence in Design* '98, J. S. Gero and F. Sudweeks (Eds.), Kluwer Academic, 1998.

M. Stefik, "Planning with Constraints (MOLGEN: Part 1 and Part 2)," *Artificial Intelligence*, *16* (2), 1981.

R. B. Stone and A. Chakrabarti (Eds.), "Engineering Applications of Representations of Function," Special Issue, *AIEDAM*, *19* (2/3), 2005.

E. Subrahmanian, R. Sriram, P. Herder, H. Christiaans, and R. Schneider, *The Role of Empirical Studies in Understanding and Supporting Engineering Design*, DUP Science, 2004.

S. Szykman and R. D. Sriram, "The NIST Design Repository Project: Project Overview and Implementation Design," NIST Internal Report, NISTIR 6926, April 2002. Available at http://www.nist.gov/manuscript-publication-search.cfm?pub_id=821838

S. Szykman, R. D. Sriram, C. Bochenek, J. W. Racz, and J. Senfaute, "Design Repositories: Engineering Design's New Knowledge Base," *IEEE Intelligent Systems Magazine*, *15* (3), 2000.

T. Tudorache, *Ontologies in Engineering: Modeling, Consistency and Use Cases, BMIR-2008-1330*, Stanford Center for Biomedical Informatics Research, 2008.

D. Ullman, *The Mechanical Design Process, 4th Edition*, McGraw-Hill, 2009.

K. Ulrich and S. Eppinger, *Product Design and Development, 4th Edition*, McGraw-Hill, 2007.

Y. Umeda and T. Tomiyama, "Functional Reasoning in Design," *IEEE Expert: Intelligent Systems and Their Applications*, *12* (2), 1997.

P. Vermaas and C. Eckert (Eds.), "Functional Descriptions in Engineering," Special Issue, *AIEDAM*, *27* (3), 2013.

W. Visser, *The Cognitive Artifacts of Designing*, Lawrence Erlbaum, 2006.

W. Visser and M. L. Maher (Eds.), "The Role of Gesture in Designing," Special Issue, *AIEDAM*, *25* (3), 2011.

R. V. Welch and J. R. Dixon, "Guiding Conceptual Design Through Behavioral Reasoning," *Research in Engineering Design*, *6*, 1994.

B. J. Wielinga and A. T. Schreiber, "Configuration-Design Problem Solving," *IEEE Expert: Intelligent Systems and Their Applications*, *12* (2), 1997.

K. L. Wood and J. L. Greer, "Function-Based Synthesis Methods in Engineering Design," *Formal Engineering Design Synthesis*, E. Antonsson and J. Cagan (Eds.), Cambridge University Press, 2001.

R. F. Woodbury and A. L. Burrow, "A Typology of Design Space Explorers," Special Issue on Design Spaces: The Explicit Representation of Spaces of Alternatives, *AIEDAM*, *20* (2), 2006.

Y. Zhang, J. K. Gershenson, and G. J. Prasad, "Product Modularity: Measures and Design Methods," *Journal of Engineering Design*, *15* (1), 2004.

Index

Printed in the United States
By Bookmasters